JAMAICA

AF767582

PEP

Mathematics

Performance Task

Practice Papers

GRADE 5

hachette LEARNING

The Publishers would like to thank the following for permission to reproduce copyright material.

Photo credits

t = top, b = below, l = left, r = right, c = centre, b/g = background

p.13: © Bruce Shippee/stock.adobe.com; p.25: © beneangulo/stock.adobe.com; p.31: © FomaA/stock.adobe.com; p.33: © Serhii/stock.adobe.com; p.41: © Julydfg/stock.adobe.com; p.54: © KatrinaEra/stock.adobe.com; p.64: © Adrian Hillman/stock.adobe.com; p.68: © DoraZett/stock.adobe.com; p.105: © New Africa/stock.adobe.com.

Acknowledgements

Although every effort has been made to ensure that website addresses are correct at time of going to press, Hachette Learning cannot be held responsible for the content of any website mentioned in this book. It is sometimes possible to find a relocated web page by typing in the address of the home page for a website in the URL window of your browser.

Hachette UK's policy is to use papers that are natural, renewable and recyclable products and made from wood grown in well-managed forests and other controlled sources. The logging and manufacturing processes are expected to conform to the environmental regulations of the country of origin.

To order, please visit www.HachetteLearning.com or contact Customer Service at education@hachette.co.uk / +44 (0)1235 827827.

ISBN: 9781036005207

First published in 2026 by Hachette Learning (a trading division of Hodder & Stoughton Limited),
An Hachette UK Company
Carmelite House
50 Victoria Embankment
London EC4Y 0DZ

www.HachetteLearning.com

The authorised representative in the EEA is Hachette Ireland, 8 Castlecourt Centre, Dublin 15, D15 XTP3, Ireland (email: info@hbgi.ie)

Impression number 10 9 8 7 6 5 4 3 2 1

Year 2030 2029 2028 2027 2026

Cover photo © GoodStudio- stock.adobe.com

Illustrations by Hyphen S.A./Konstantina Kyratzidou

Typeset by Hyphen S.A.

Printed and bound in Great Britain by Bell & Bain Ltd, Glasgow

A catalogue record for this title is available from the British Library.

FSC www.fsc.org
MIX
Paper | Supporting responsible forestry
FSC™ C104740

Contents

How to use this book

Think about successful athletes. Athletes learn about their sport, and they practise and train hard to become excellent performers.

You can be a maths-athlete who performs well in tasks by learning how to do them and then training yourself to answer different types of questions.

You are going to work through **Practice paper 1** with your teacher. This will help you to:

- Read questions carefully and work out what type of answer to give.
- Understand why it is important to show how you work out your answer.
- Learn how you can get good results even if your answer is not completely correct.
- Understand what the different ratings mean and see how you can improve your rating.
- Reflect on your own work to help you see what you did well and what you can improve in the next task.

Each task in this book has three supporting features.

Before you start …

This section helps you think carefully about questions and what it is you are being asked to do. There are different activities for each task.

Hints

These are given in the task itself to help you think about the questions and how you should answer them. As you progress, there will be fewer hints in the tasks.

Reflection

This section encourages you to think about your own learning and how you can improve on it.

Practice paper 1 also has a section called **How do you know how well you did?**

In this section you will see possible answers to the questions and what rating they were given. The aim of this section is to show you what you need to do to get the best possible results.

Practice paper 1

Always check that you have everything you need before you start.

This task has 3 parts:

- Part 1 has questions 1 and 2.
- Part 2 has questions 3 and 4.
- Part 3 has question 5.

You will need a calculator, a pencil and an eraser.

Answer all the questions in each part in the spaces provided.

Remember to always show your working out.

When you do a performance task, it is helpful to read through all the questions before you start so you can think about what you need to do and plan your time.

→ Before you start …

The instruction words in each question are very important. Here are some of the instruction words you might find in a performance task.

Find	Determine	Decide	Write	Draw
Show	Complete	Suggest	Choose	Explain
Describe	Calculate	Work out	Give reasons	Match

Start by **reading** all the questions. **Circle** or highlight all the **instruction words**.

Now, **tick (✓)** questions that look like they have only one short answer.

Next **find the underlined instructions** in the task.

What do you have to do when you answer these questions? **Share your ideas** with your partner.

Setting up an art club

A community organisation plans to set up an art club at the community centre. They will offer classes for children and adults, and they will have an exhibition at the end of each term.

The art club committee has divided into teams. Each team will handle different aspects: Team A organizes the room, Team B plans activities, and Team C analyses the exhibition data. You will help with tasks from all three teams.

Part 1

Organising the art room

Question 1

The art club has been given a room in the community centre. The plan of the room is shown here.

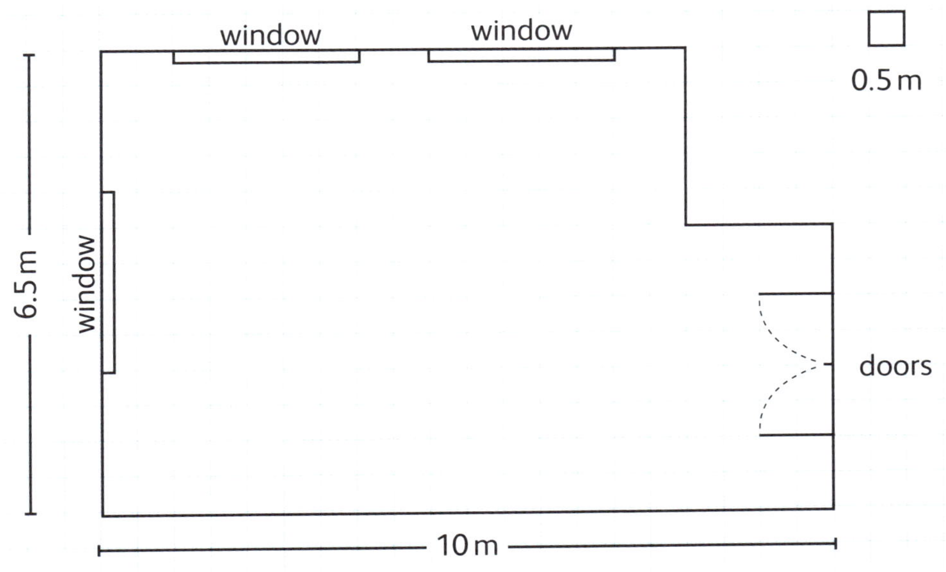

Use the scale on the grid to help you work out the missing measurements on the plan.

a Fill in the missing number in each sentence.

- There are ______ windows in the room.
- There are ______ doors.
- The room is ______ m long.
- Each door is ______ m wide.
- The wall with two windows on it is ______ m long.

You can draw on the plan. You can also divide it into shapes you know.

b Calculate the floor area of the room. Show all the working you do.

Start by thinking about where you will put the different items. Use the information above to help you decide.

c The committee has to decide where to put the tables and stools, the shelves, the supply cupboard and the noticeboard. They agree on the following conditions:

- The students should have good natural light when they are working.
- The stools will fit under the tables, but there should be space around each table so the teacher can move around each group as they work.
- The shelves will be used to store finished artwork.
- The supply cupboard should be easy to get to.
- The noticeboard should be where everyone coming into the art room can see it easily.

How would you arrange the room to fit in these pieces of furniture?

Remember to use the correct scale for each item.

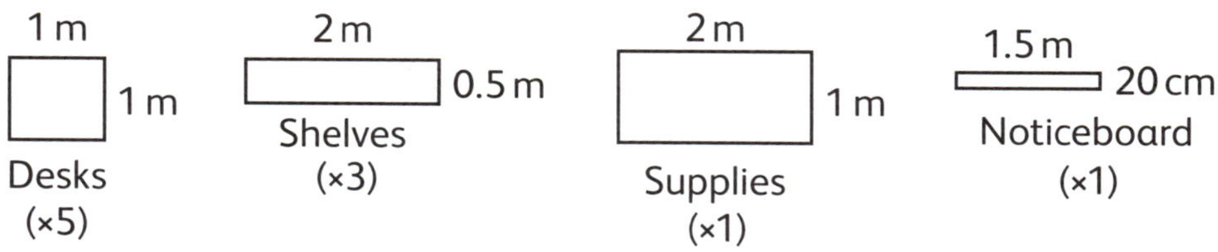

Think about what you decided and try to explain clearly why you put the items in different places.

i Draw your arrangement <u>on the plan.</u>

ii <u>Complete</u> these sentences.

- I put the chairs and tables there because ______________________________.
- I think this is a good place for the shelves because ______________________________.
- I put the noticeboard in that position because ______________________________.
- I put the supply cupboard there so that ______________________________.

Question 2

Highlight the prices on the images.

The students in the art club plan to paint the room. They estimate that they will need 40 litres of paint and 6 paint rollers to do the job. They find these prices online:

$4 994.80 | $20 877.92 | $2 667.88

Your answer must include the number and size of the tins as well as the total cost.

a How much will it cost to buy 6 rollers?

b What is the cheapest combination of tins of paint that the committee can buy to make sure they have 40 litres? Explain your decision.

When you get a discount, the store takes off that amount of money from your total.

c The final amount at the hardware store comes to $52 086.32. The store gives the committee a 15% discount because this is a community project. How much will they pay?

Part 2

Planning activities

Question 3

Ms Campbell is in charge of arranging and scheduling different activities. She arranges these activities for September.

September in the studio

Activity	Class dates and times	
Design your own fabric	Saturday 2nd	9 a.m. to 5 p.m.
Learn pencil sketching	Wednesday 6th Friday 8th	6:30 p.m. to 8:30 p.m. 8 a.m. to 10 a.m.
Cartooning on computer	Friday 8th Saturday 9th	3 p.m. to 6 p.m. 10 a.m. to 12 noon
Oil painting	Saturday 16th / 23rd	11 a.m. to 3 p.m.
Pastels and watercolours	Every Thursday	5 p.m. to 7 p.m.
Meet and draw a live parrot	Saturday 30th	3 p.m. to 5 p.m.

Some activities are offered on more than one day.

a How many different activities are offered?

____

b Malia can only do classes in the evening. What activities could she take?

____

____

Think about how we write time to show it is in the evening.

c On which dates could you learn to draw cartoons using a computer?

____

____

Remember, this is all happening in September.

Find the information on the poster.

d Marc wants to meet a parrot and learn to draw it. When can he do that?

e How long does the fabric design class take?

Will a number line help you work this out?

Question 4

Draw and label a sketch or number line to show the problem.

a Seventeen people sign up for the fabric design class. Each person will receive 0.6 m of white fabric to work on. If the roll of fabric is 12 m long, how much will be left after giving each student their piece of fabric?

What does fewer mean?

b 49 people sign up for the pencil sketching classes. The Wednesday class has five fewer people than the Friday class. How many people are in each class?

You need to use your answers from part b to work this out.

c The teacher wants to arrange the pencil boxes on the supply shelf. She has boxes of HB, 3B, and 5B pencils. Draw two different ways she could arrange them so they're easy to find.

Remember, the people sit facing the table. This will help you work out left and right.

d Maria is teaching the class on 30 September. She organises the tables and puts a name label on each place to show people where to sit. This is her seating plan with some of the names filled in.

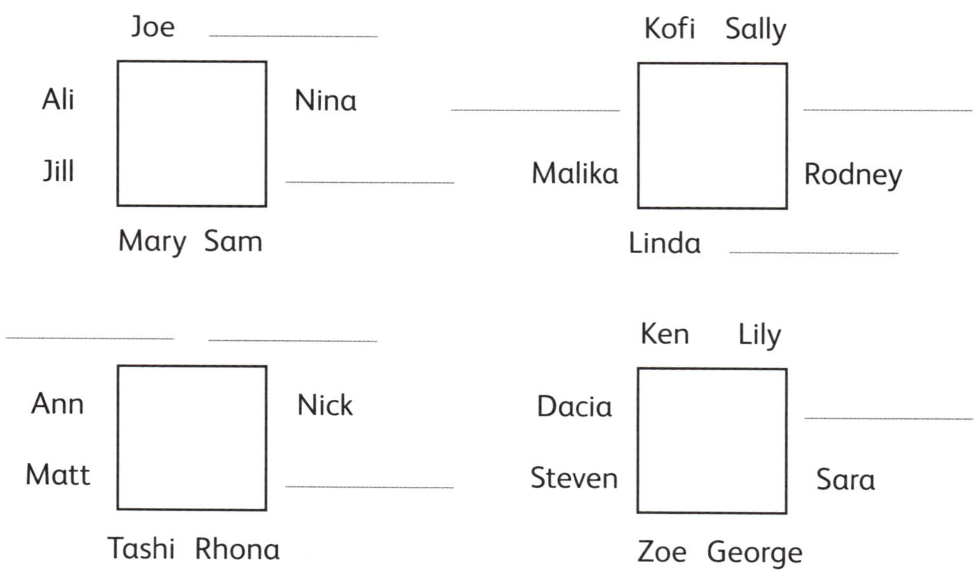

Use this information to fill in the missing names on the plan.

- Tom is on Rodney's left.
- Lydia is to the right of Sara.
- Mike is on Rhona's right, at the same table.
- Ciara is opposite Sam.
- Monica is behind Sam.
- Julie is next to Nina and Sam.
- Ruth is next to Mary, but at another table.
- Tay is between Kofi and Malika.

e How many people are booked to attend the class?

____

Look at the arrangement of people and think about a sensible place for the parrot. Remember to mark it clearly.

f Where would you put the parrot so that everyone can see it and draw it? Mark the position on the seating plan diagram with an X.

Part 3

Ticket sales and attendance

Question 5

At the end of each term, the art club has an exhibition of student work. The members sell tickets to raise money for art supplies and equipment.

Ticket prices are:

Adults	\$350
12–18 years	\$140
Under 12	Free

Suresh sells tickets to the exhibition. The first ticket on the roll is number 4 732.

The ticket numbers increase as he sells them.

a Suresh sold 320 tickets in the first week. What was the last ticket number he sold?

To find the size of each face, multiply the length by the width, the length by the height or the width by the height.

b A ticket roll is kept in a rectangular box with a length of 20 cm, a width of 10 cm and a height of 6 cm. Which face is the biggest: top, front or side? Show your working.

You need to do more than one step here. Show all your working.

c In total, $\frac{3}{5}$ of the tickets were adult tickets. How much money did Suresh take in ticket sales?

Read the key and the labels on the graph carefully.

d The exhibition was open from 8:30 a.m. till 5:30 p.m. on Saturday and Sunday. The graph shows the numbers of people attending the exhibition at different times on Saturday and Sunday.

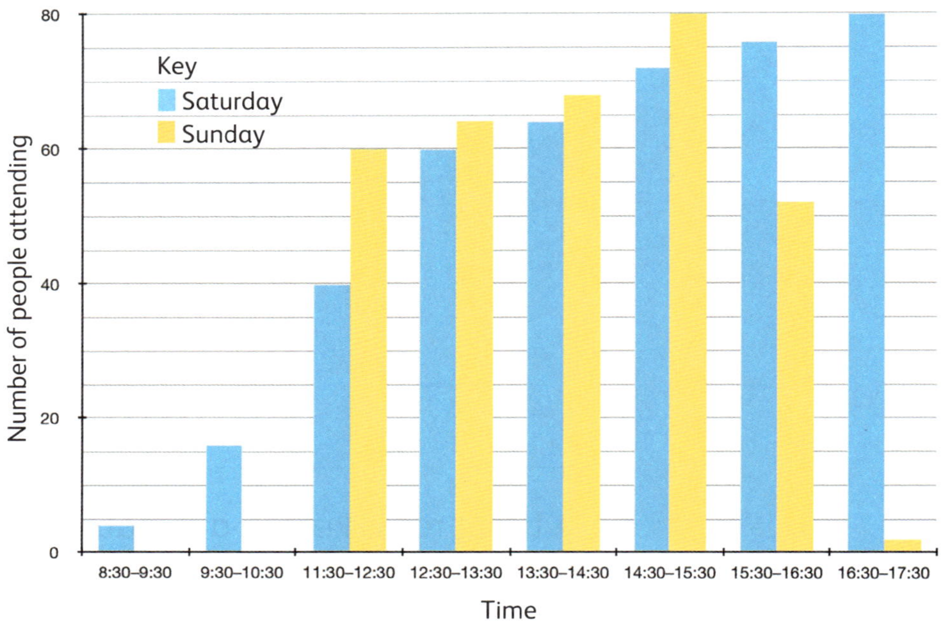

Think about the busy and not so busy times.

Analyse the graph and write suggestions to help the committee decide what opening hours they should have for the next term's exhibition. Give reasons for your suggestions.

Reflection

Go back through your own answers to the questions in this task.

- Tick the boxes for skills you used successfully in this task:

  - [ ] Solving problems with multiple steps
  - [ ] Explaining my reasoning clearly
  - [ ] Showing my working step-by-step
  - [ ] Other ____________________

- Which ones did you struggle with? What made them difficult for you?

  - [ ] I didn't understand what to do
  - [ ] The calculations were too hard
  - [ ] There were too many steps
  - [ ] I couldn't find the information I needed
  - [ ] I ran out of time
  - [ ] Other ____________________

- Based on this task, I will:

  Practice more with:

  ____

  Next time I will remember to:

  ____

  One thing I did really well was:

  ____

Practice paper 1 Self-evaluation

How do you know how well you did?

Now that you have worked through the task, you are going to look at the answers that some students gave to see how they were rated. Pay attention to what students did well and think about why some answers are better than others.

Once you have read all the answers, you can try to rate your own answers from 1 to 4 using this scale.

1 Beginning	2 Developing	3 Proficient	4 Highly proficient
Your answers tell the teacher that you do not understand the work and that you need further support.	Your answers show that you understand some of the work but that you need help in certain topics.	Your answers show that you understand most of the work.	Your answers show that you understand all of the work very well.

Question 1

a Fill in the missing number in each sentence.

• There are 3 windows in the room. ✓ • There are 2 doors. ✓ • The room is 6.5 m long. ✗ • Each door is $\frac{1}{2}$ m wide. ✗ • The wall with two windows on it is 10 m long. ✗	• There are 3 windows in the room. ✓ • There are 2 doors. ✓ • The room is 10 m long. ✓ • Each door is 2 m wide. ✗ • The wall with two windows on it is 8 m long. ✓
2 Developing *The student does not read the plan well and is not able to work with the scale. Also, the student does not know the difference between length and width.*	3 Proficient *The student has answered 4 out of the 5 questions correctly. Answers show they can read the plan and use the scale.*

b Calculate the floor area of the room. Show all the working you do.

1 Beginning	4 Highly proficient
$6.5 \times 10 = 65$	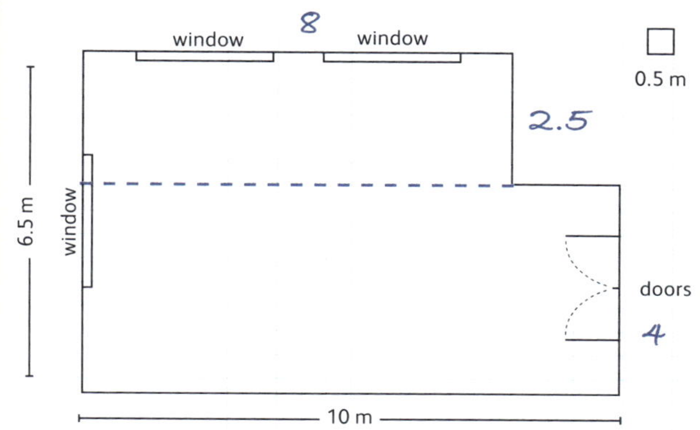 $8 \times 2.5\ m = 20\ m^2$ $4 \times 10\ m = 40\ m^2$ $20\ m^2 + 40\ m^2 = 60\ m^2$
The student's answer is incorrect and has no units. *Their working shows they do not understand how to find the area of a compound shape.*	*The student has shown clearly how they divided the shape into two rectangles, and has worked out the area of each correctly and added them to find the total area.*

c How would you arrange the room to fit in these pieces of furniture?

i Draw your arrangement on the plan.

ii Complete these sentences.

- I put the chairs and tables there because …
- I think this is a good place for the shelves because …
- I put the noticeboard in that position because …
- I put the supply cupboard there so that …

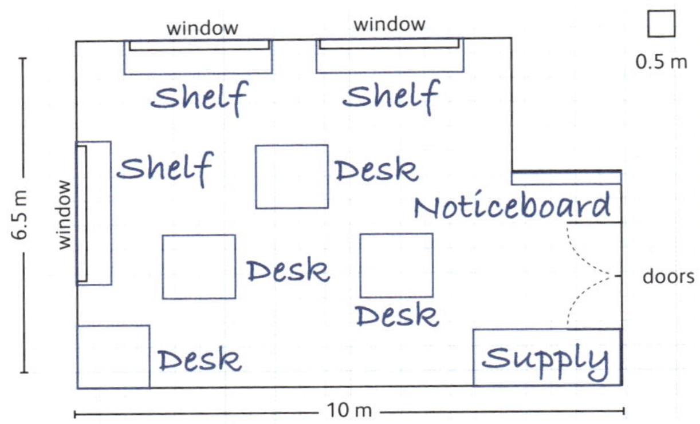

- I put the chairs and tables there because they fit the space.
- I think this is a good place for the shelves because it is out of the way.
- I put the noticeboard in that position because it's near where people come in.
- I put the supply cupboard there so that the teacher can get stuff when she comes in.

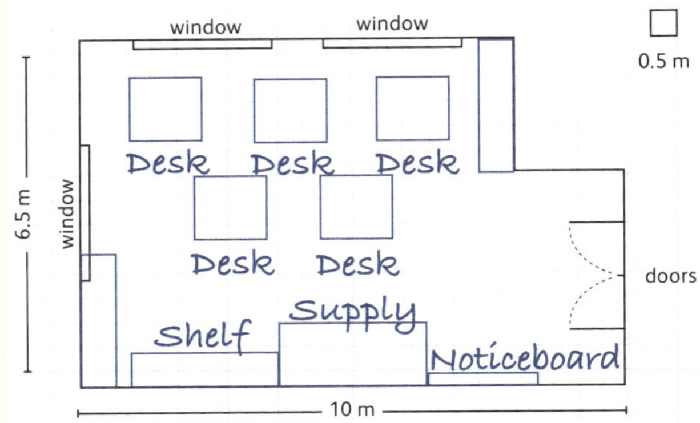

- I put the chairs and tables there because they are near the windows and not in the way of the door.
- I think this is a good place for the shelves because they are not in the way.
- I put the noticeboard in that position because when people walk into the room they will have to walk past it.
- I put the supply cupboard there so that there's nothing in front of it.

2 Developing	3 Proficient
The student has shown the items on the plan and labelled them, but some choices are not sensible (noticeboard is behind the open door, shelves under the windows mean less light for the work desks). *The supply cupboard is behind the door so it's hard to get to.* *Only four desks are shown and one is in the corner.* *Reasons are given, but not all are sensible.*	*The student has included all the items on the plan in sensible positions, but has not labelled them all.* *They have given sensible reasons for their choices.*

Question 2

a How much will it cost to buy 6 rollers?

b What is the cheapest combination of tins of paint that the committee can buy to make sure they have 40 litres? Explain your decision.

c The final amount at the hardware store comes to $52 086.32. The store gives the committee a 15% discount because this is a community project. How much will they pay?

a 2 × $2667.88 = $5335.76 **b** Cheapest is 2 × $20877.92 + $4994.80 = $46750.64 **c** 15% of $52086.32 = $7812.948 and $52086.32 − $7812.948 = $44273.372 The cost is $44273.73	**a** There are three rollers in a pack, so they need two packs. 2 × $2667.88 = $5335.76 **b** The cheapest combination is two 19-litre tins (which gives 38 litres) and an additional 3.8-litre tin to get the 40 they need. The cost for this is 2 × $20877.92 + $4994.80 = $46750.64 **c** 85% of the total is 0.85 $52086.32 × 0.85 = $44273.34 (rounded to 2dp)
③ Proficient	④ Highly proficient
This student has written number sentences and shown the steps in their working. *In part (b), they worked out the right amount but have not fully answered the question because they have not said how many tins of each size they need.* *In part (c), they have made a careless error (writing 73 instead of 37) when they wrote the answer.*	*This student has laid out all their working very clearly and they have checked their answers to make sure they are all correct.* *In part (b), they have explained that they need to buy slightly more paint than needed because of the size of the tins.*

Question 3

a How many different activities are offered?

b Malia can only do classes in the evening. What activities could she take?

c On which dates could you learn to draw cartoons using a computer?

d Marc wants to meet a parrot and learn to draw it. When can he do that?

e How long does the fabric design class take?

a six **b** 2 classes **c** on the 8th and 9th **d** on Saturday from 3 p.m. to 5 p.m. **e** seven hours	**a** six classes **b** pencil sketching or pastels and watercolours **c** 8th and 9th **d** at 3 p.m on 30 Saturday **e** it takes eight hours 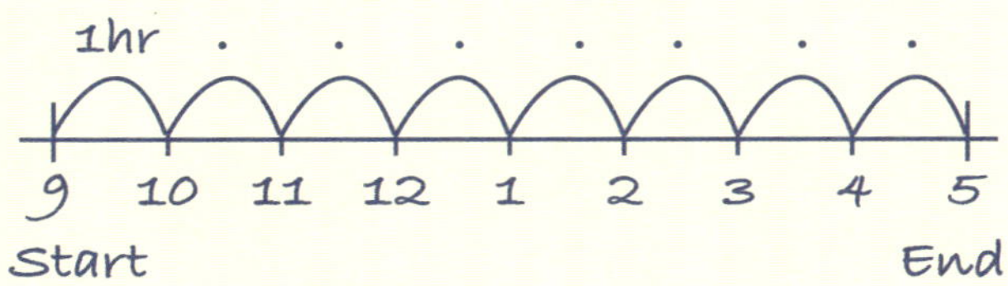
② Developing *The student has tried to answer all the questions and seems to have read the poster correctly, but they have not written the names of the classes in full in part (b) or shown how they worked to find the time in part (e). If they had shown their working, they would get credit even though the answer was not completely correct.*	③ Proficient *The student has answered all the questions correctly. This shows they have read and understood the poster.* *In part (c), they should give the month and day as well as the date.* *In part (e), the number line shows they understand how to work out how long something takes.*

Question 4

a Seventeen people sign up for the fabric design class. Each person will receive 0.6 m of white fabric to work on. If the roll of fabric is 12 m long, how much will be left after giving each student their piece of fabric?

b 49 people sign up for the pencil sketching classes. The Wednesday class has five fewer people than the Friday class. How many people are in each class?

c Each person who signed up for the pencil sketching class is given three different pencils (HB, 3B and 5B). The pencils come in boxes of 12. How many boxes of pencils will the teacher need altogether?

a 17 × 0.6 m = 10.2 m b Try 25 and 24 ✗ not 5 more 21 and 28 ✗ not 5 more 23 and 28 One class has 23 and the other has 28 c 49 × 3 = 147 pencils 147 ÷ 12 − 12.25 boxes of pencils	a They need 17 × 0.6 = 10.2 m They have 12 m. 12 m − 10.2 m = 1.8 m left b 49 − 5 leaves 44 $\frac{1}{2}$ of 44 = 22; so one class will have 22 and the other will have 22 + 5 = 27 c There are three types of pencils and the teacher needs 49 of each type. 4 × 12 = 48 pencils. That is not enough. 5 × 12 = 60 pencils. This means she needs 5 boxes of each type of pencil. 15 boxes in total.
2 Developing *The student has shown their working but they have not managed to do the calculations correctly.* *In part (a), they forgot to subtract to get the amount left.* *In part (b), the answers don't add up to 49.* *In part (c), they forgot you cannot buy part of a box.*	4 Highly proficient *The student has shown all of their working and all of their answers are correct.*

d Use this information to fill in the missing names on the plan.

e How many people are booked to attend the class?

f Where would you put the parrot so that everyone can see it and draw it? Mark the position on the seating plan diagram with an X.

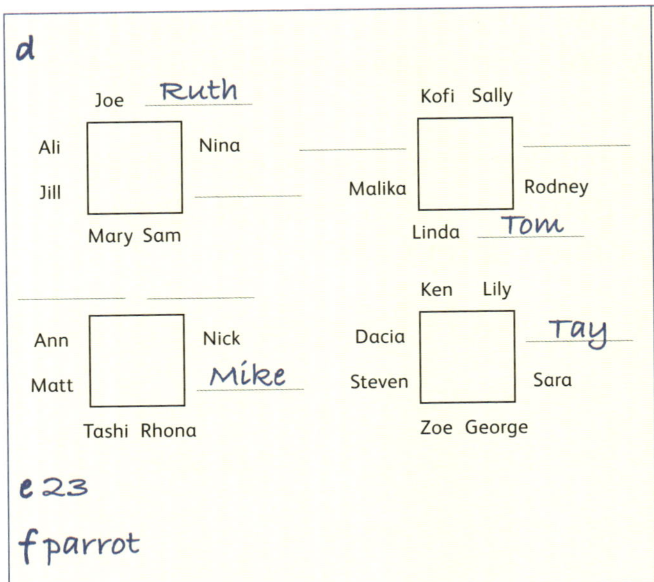

❶ Beginning	❹ Highly proficient
The student has not filled in the plan properly and all the names they have written are in the wrong place. *In part (e), they counted the names on the plan already (23).* *They did not show where the parrot should go on the plan.*	*The student has shown all the names on the plan in the correct places and has put a circle to show one place is empty.* *They have done a calculation to work out the number of people and subtracted the empty desk to get the right answer.* *They have a good position for the parrot in the middle of the room.*

Question 5

a Suresh sold 320 tickets in the first week. What was the last ticket number he sold?

b After Suresh sold all the tickets for this term, the first ticket on the roll was 5 472. How many tickets were sold altogether?

c $\frac{3}{5}$ of the tickets were adult tickets. How much money did Suresh take in ticket sales?

a 5050 b $5472 - 4732 = 742$ tickets c $\frac{1}{5}$ is $740 \div 5 = 148$ $\frac{3}{5}$ is $3 \times 148 = 444$ adults $444 \times 350 = \$155400$ $\frac{2}{5}$ is $2 \times 148 = 296$ children $296 \times 140 = \$41440$	a $4732 + 320 = 5052$ b $5472 - 4732 = 740$ tickets c $740 \div 5 = 148 = \frac{1}{5}$ $3 \times 148 = 444$ adults; $444 \times 350 = \$155400$ $2 \times 148 = 296$ children; $296 \times 140 = \$41400$ $\$155400 + \$41440 = \$196400$
2 Developing	3 Proficient
The student has shown some working out but they have not managed to get all the answers correct. *In part (a), they got close to the answer, but they didn't show their working so they cannot get full credit.* *In part (b), they miscalculated, but they get credit for showing the right calculation.* *In part (c), they forgot to add the totals together to get the answer.*	*The student showed all their working and they got the correct answers in parts (a) and (b).* *In part (c), they made two careless errors so they did not get the right answer, but they get credit because they worked correctly.*

d Analyse the graph and write suggestions to help the committee decide what opening hours they should have for the next term's exhibition. Give reasons for your suggestions.

I think they should have the same opening hours because there were people at all the times.	I think next year they should consider opening on Saturday from 11:30 a.m. to 18:00 p.m. and on Sunday from 11:30 a.m. to 16.30 p.m. On Saturday and Sunday, very few people attended before 11:30 a.m., so you would not lose many visitors if you opened a bit later. On Sunday, very few people attended after 16:30 p.m., so you could close earlier on a Sunday (maybe people have to get to church or prepare for Monday morning). I also suggest that you stay open a half-hour longer on Saturday as early evening was the busiest time and people can attend later on Saturdays.
② Developing *The student has made a suggestion and given a reason but they have not taken the number of people at different times into account.*	④ Highly proficient *The student has given their suggested opening hours very clearly. Then they have given detailed reasons for each of the changes that they have suggested.*

Practice paper 2

This task has 3 parts:

- Part 1 has questions 1 and 2.
- Part 2 has questions 3 and 4.
- Part 3 has question 5.

Answer all of the questions by writing or drawing in the space provided. Remember to show any working out that you do.

→ Before you start …

Complete these four multiple choice items to remind you about the types of questions you might get in a performance task and how you can think about and answer each type.

1 The height of the door is 200 cm. What is the height of the door in metres?

To answer this question, I would:

(A) multiply by 100.

(B) divide by 100.

(C) multiply by 10.

(D) divide by 10.

Lemonade Recipe (for 10 servings):

- 175 mL of white sugar
- 1.5 L of water
- 250 mL lemon juice

2 Read the list of ingredients you need to make ten 250-millilitre servings of lemonade.
How many millilitres of white sugar is needed to make fifteen 250-millilitre servings of lemonade?

To answer this question, I would:

(A) subtract 10 from 15, then multiply 175 by the difference.

(B) divide 15 by 10, then multiply 175 by the quotient.

(C) add 15 and 10, then multiply 175 by the sum.

(D) multiply 175 by 15.

Use the scenario below to answer questions 3 and 4.

Jayden and his brother arrive at the zoo at 10 a.m. It took Jayden 30 minutes to eat breakfast, 45 minutes to do his chores and an additional 35 minutes to get ready. It takes 17 minutes to travel by car from home to the zoo.

3 What time does Jayden leave home?

To answer this question, I would:

(A) count backwards 17 minutes from 10 a.m.

(B) find the difference between 10 and 17.

(C) guess a random time before 10 a.m.

(D) count backwards 35 minutes from 10 a.m.

4 About what time does Jayden need to start his day so that he can be ready to leave home and get to the zoo on time?

To answer this question, I would:

(A) add up the amount of time Jayden takes to complete his tasks and get ready, then count backwards from the departure time.

(B) add up the amount of time Jayden takes to complete his tasks and get ready, then count backwards from the time he arrives at the zoo.

(C) add up the travel time and the amount of time Jayden takes to complete his tasks and get ready, then count backwards from the departure time.

(D) estimate the amount of time Jayden takes to complete his tasks, add an hour to that, then subtract this estimated time from the departure time.

Measure me this! Measure me that!

Shanoy is 11 years old. She loves sports, computers and baking with her family on the weekend. This weekend is no different. Today is a busy day for Shanoy because she is going for a health check and then some of her friends are visiting in the evening.

Shanoy needs your help to:

- Record and make sense of her health check results.
- Explain what your BMI can tell you.
- Work out what ingredients she needs.
- Plan her time.

Part 1

The health check

Question 1

Shanoy's mom takes her to the clinic for a health check-up. At the clinic, the nurse checks her temperature and measures her mass and her height.

Shanoy's temperature

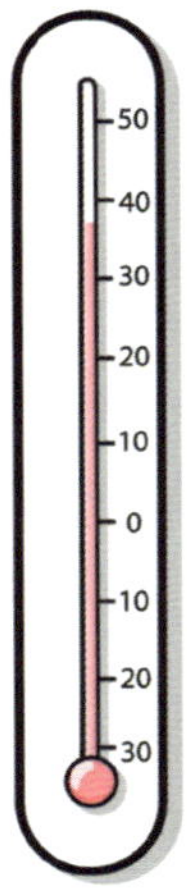

Shanoy's height (in cm)

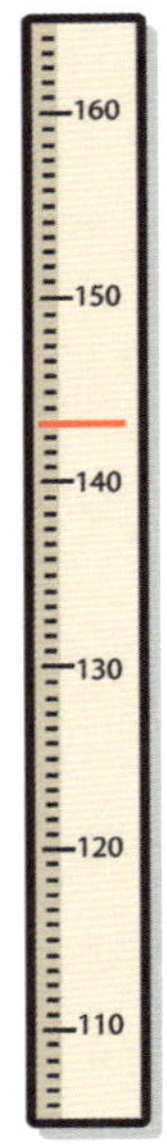

> Make sure you know which instrument is used to measure temperature, mass and height. Use the units as a guide.

a The thermometer shows Shanoy's temperature. Record this using the correct units in the space below the thermometer.

b Shanoy weighs $42\frac{1}{2}$ kg. Draw a line on the scale to show her mass.

c Shanoy's bathroom scale is digital. How would her scale display a weight of $42\frac{1}{2}$ kg?

> Think about how digital displays show numbers.

____

Why?

____

> Shanoy's height is shown by the red line.

d Shanoy's height is marked on the cm scale. Mark and label these heights on the scale:

- Your own height
- The average height for an 11-year-old (approximately 145 cm)
- A height that would be very tall for an 11-year-old.

Body Mass Index (BMI)

Body Mass Index is a simple calculation that compares a person's height and weight. The formula is BMI = kg/m^2 where kg is a person's weight in kilograms and m^2 is their height in metres squared. Health workers use BMI together with other measurements to determine whether a person is under or over the normal weight for their age and height.

Question 2

Write your answer here, correct to one decimal place.

a Complete the table to determine Shanoy's Body Mass Index (BMI).

Shanoy's weight (in kg)	Shanoy's height (in metres)	Find the square of Shanoy's height	BMI = (weight in kg) ÷ (square of height)

The nurse at the clinic shows Shanoy this chart with her BMI shown on it as a vertical line.

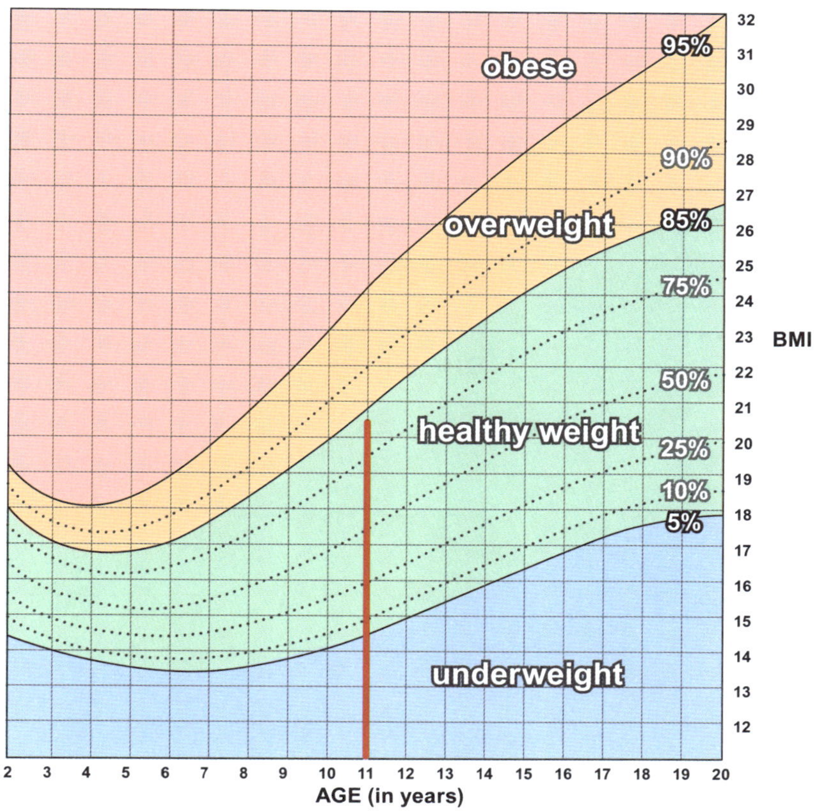

Read the labels on the axes and on the graph itself before you start.

b What does Shanoy's BMI say about her health? Use the chart to help you complete this report from the health worker.

Report

Shanoy has a similar weight to almost ______ % of girls her age. Her weight is classified as ______. Girls of this age who are of a healthy weight can have a BMI between ______ and ______. A BMI above ______ would suggest that an 11-year-old girl was obese. A BMI below ______ would suggest that an 11-year-old girl was underweight. Shanoy should continue to eat healthy foods and exercise regularly to enjoy good health.

You should remember from Science that muscle tissue is heavier than fat.

c BMI is just one measure of a person's health. It is possible for a healthy athlete to have a high BMI, showing them to be overweight. A healthy child may have a low BMI, showing him or her to be underweight. Suggest two reasons for this.

______

______

Refer to the graph in your answer.

d Shanoy's sister, Janelle, is 15. She is 1.62 m tall and she weighs 63 kg. At her health check, Janelle asks the nurse if she should be on a weight loss diet because she thinks she is very overweight. Imagine you are the nurse at the clinic. How would you answer Janelle's question?

______

______

______

______

______

Part 2

Sweet treats

Shanoy has a basic muffin recipe that makes 12 muffins.

Shanoy's delicious muffin treat

Muffin recipe

2 eggs, large, refrigerated
2 cups all-purpose flour
2 tsp baking powder
½ cup granulated sugar
½ tsp salt
½ tbsp sugar, coarse
½ cup butter, unsalted
¾ cup milk

Question 3

Shanoy receives a WhatsApp message that some of her friends are planning to visit so they can watch a movie together. Shanoy decides she will need to make 18 muffins.

The diagram shows how 12 muffins fit in a muffin tray:

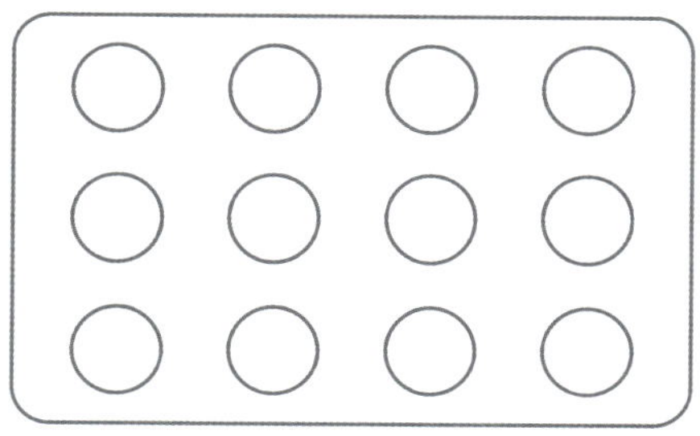

Draw two different ways Shanoy could arrange 18 muffins in trays.

Janelle eats $\frac{1}{9}$ of the muffins once they are ready. Shanoy says there are now enough muffins for her and her friends to have three each, with one left over. If Shanoy and each friend eat two muffins, there will be six left over. How many friends are coming over?

Question 4

While they are baking, Shanoy's mom bakes some goodies to sell at the church fair. This is what she bakes:

a How do you think Shanoy's mom worked out what to charge for each item? Explain your reasoning.

b The deacon and her friend buy $\frac{1}{4}$ of the gizzadas, $\frac{1}{3}$ of the cookies and $\frac{1}{6}$ of the muffins. They share the costs equally. How much will they each pay?

Part 3

Games and a movie

Question 5

Shanoy is expecting her friends to arrive at 6:45 p.m.

It takes 40 minutes to bake the muffins and Shanoy wants them to be ready half an hour before so that they can cool.

Remember, hops on a number line are useful when you have to solve problems involving time.

a At what time should Shanoy put the muffins in the oven to be ready half an hour before her friends arrive?

Shanoy's friends arrive together. The clock shows the time her friends arrived.

Draw a sketch or model to help you here.

b What time of the evening did the friends arrive at Shanoy's home?

c Shanoy's friends arrived 10 minutes before the start of the movie. What time did the movie start?

d When the movie ended, it was 9:33 p.m. How long was the movie?

e Shanoy and her friends played board games for 45 minutes after the movie ended. After the game was up, they left to go home.

i Write the time that her friends left to go home in 24-hour notation.

ii Draw a timeline to show how long Shanoy's friends stayed at her house altogether.

Label your timeline with both the times and the events.

Reflection

Looking closely at your mistakes can help you learn and improve your work.

Think about any questions or parts of questions that you got wrong.

- Did you understand the question and know what to do? If not, what can you do about this?

- Did you make any mistakes in your calculations? If so, what can you do to avoid these mistakes in future?

Practice paper 3

This task has 4 parts:

- Part 1 has questions 1 and 2.
- Part 2 has question 3.
- Part 3 has question 4.
- Part 4 has question 5.

Answer all the questions.

Write or draw your answer in the spaces provided.

Remember to show all your working.

→ Before you start ...

In this task, you have to find information from different sources.

Find these sources of information:

- facts on a website
- race times in a table
- a set of raw data.

How will you use each source?

Write your ideas here:

Website: ________________________________

Race times: ______________________________

Raw data: ________________________________

Athletics

Athletics is an exciting sport and Jamaicans are very proud of their athletes. In this task, you are going to read about international events and answer questions about time and distance. Then you will use your skills to organise information, make decisions about winning times, plan a sports area and draw and interpret graphs using data about different events.

Part 1

Field events include javelin, discus, shot put and long jump.

What are the events?

Question 1

Read this information about the Diamond League athletics events before you answer the questions.

About Menu

The Diamond League is an annual series of athletics meetings that takes place in 14 different countries throughout the year. The series started in 2010 and it comprises both track and field events. Events include sprints (100 m, 200 m and 400 m), hurdles, relays and longer races (800 m, 1 500 m, 3 000 m and 5 000 m). Other events are shot put, pole vault, high jump, triple jump, discus and javelin. In 2023, 18 elite Jamaican athletes took part in the events. Shericka Jackson did very well in both the 100 m and 200 m sprints, and Natoya Goule-Topping dominated in the 800 m event. Tajay Gale did us proud in the long jump and Rasheed Broadbell shone in the hurdles. The men's 100 m final included Ackeem Blake, Yohan Blake and Kishane Thompson.

a Name any **two (2) field** events in the series.

____

b Identify **one (1) track** event in which a Jamaican athlete participated.

____

c Shericka Jackson won the 200 m event in Morocco with a time of 21.91 seconds. Later in the year, she ran the same event in Switzerland with a time of 21.48 seconds. How much faster did she run in the Swiss event?

21.91 – 21.48 = ____________

Think about how many seconds are in a minute and how many minutes are in an hour.

d Natoya Goule-Toppin ran 800 m in 1:58.23. Explain what this time means.

____

____

World record: 1:53.28
Natoya's time: 1:58.23
What is the difference in time? Use a number line if it helps.

e The world record for the 800 m is 1:53.28. How much faster would Natoya Goule-Topping have to run this distance to beat this record?

____

Read the exchange rate carefully.

f The prize money for the overall winner is US \$30 000. Use a rate of US \$10 = J\$1 549 to convert this amount to Jamaican dollars.

Question 2

One of the Diamond League athletics meetings takes place in Doha, in Qatar. The map shows four routes that you can fly from Jamaica to Doha.

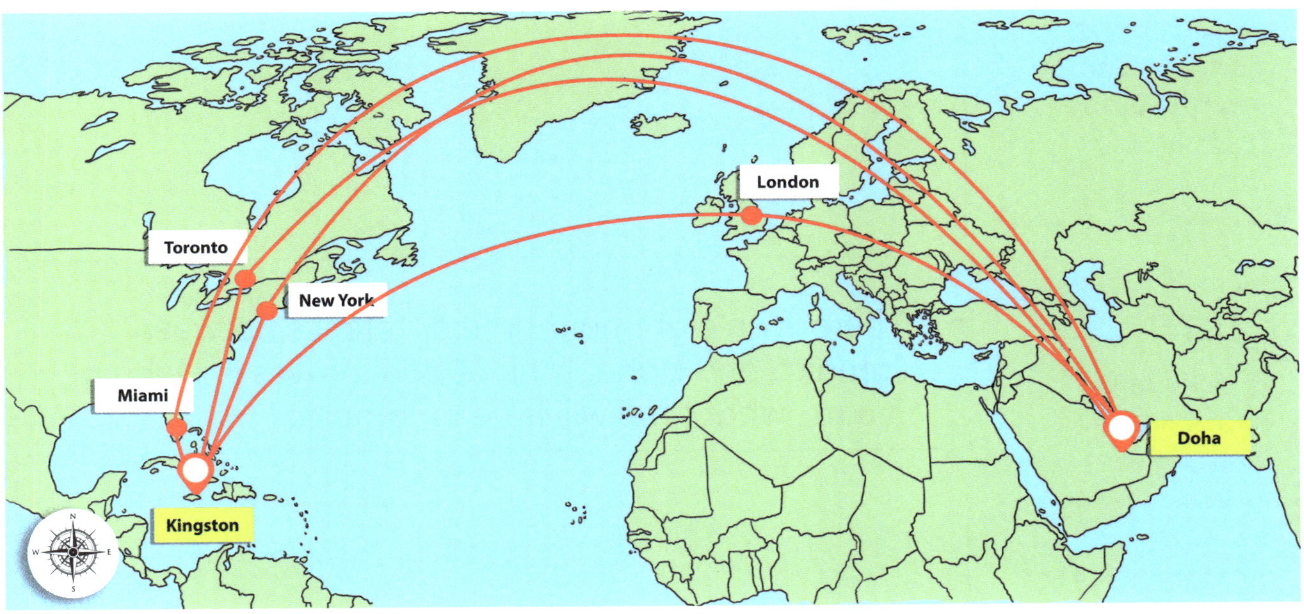

a Which route is the shortest in distance? Explain how you decided.

____

____

Make a sketch of the main compass directions to help you.

b Complete the sentences by circling the correct words in bold.

- I start my journey in Kingston, Jamaica.
- I fly **north/east/south/west** to Miami.
- Then I travel across the ocean to **Kingston/Doha/London**.

c The flight that goes through Toronto takes the longest time to get to Doha. Suggest one reason for this.

____

____

When you get off a plane and spend time at the airport before getting onto another plane, it is called a layover.

d One athlete travels from Jamaica to London and then to Doha. The journey takes 19 hours and 5 minutes. Another athlete travels from Jamaica to New York and then to Doha. The journey takes 21 hours. How much shorter is the trip through London?

If Doha is 8 hours ahead, what do we subtract from 3a.m.?

e There is an eight-hour time difference between Jamaica and Doha. If an athlete arrives in Doha at 3 a.m., what time will it be in Jamaica?

Part 2

Sports day

Halloway Primary School is having their sports day. All the students are excited as they will be representing their houses in various events and activities on this special day.

Question 3

First: Find the speed: $50 \div 9$ = ____

Then: Change km to m: 2.3×1000 = ____

Last: Find the time: ____ $\div$ ____ = ____

a For the first track event of the day, students are to run a marathon – cross country race.

i If one student runs 50 m in nine seconds, how long will it take to run 2.3 km at the same speed?

60 seconds = 1 minute. Divide your answer by 60, then write as a mixed number and simplify.

ii Rewrite your answer in minutes as a mixed number in its lowest form.

Look for the smallest number in the 'time' column for boys.

b Their times for the top four runners in the semi-final are given in the table below:

Names	Time (seconds)
BOYS RACE	
Mark Weller	18:35
John Dyer	18:01
Ian Grey	19:00
Nayor Green	18:16
GIRLS RACE	
Lisa Brown	19:21
Nadia Drummond	19:25
Rachel Hue	20:14
Iesha Keller	20:14

i Which boy ran the fastest time?

____

ii Which girl finished second?

____

iii What is the difference between the fastest and slowest girl's time in this race?

A decimal with two places has: Tenths (first place): 0.5 Hundredths (second place): 0.05 Five-hundredths: 0.05

To find new times, subtract 0.05 from each time.

c The organisers decide to remove five-hundredth of a second from each runner's time because the conditions were very windy.

i Write new times for each student and rank them from 1st to 4th. The first one has been done for you.

Names	Time (seconds)	Students' Position
BOYS		
Mark Weller	18:30	1st
John Dyer		
Ian Grey		
Nayor Green		
GIRLS		
Lisa Brown		
Nadia Drummond		
Rachel Hue		
Iesha Keller		

ii Will removing five-hundredths of a second from each runner change their positions in the race? Explain your answer.

____

____

____

$2\frac{1}{2}$ laps means 2 whole laps + half a lap

Whole laps: 2 × 200 = ____ m

Half lap: 200 ÷ 2 = ____ m

Total: ____ + ____ = ____ m

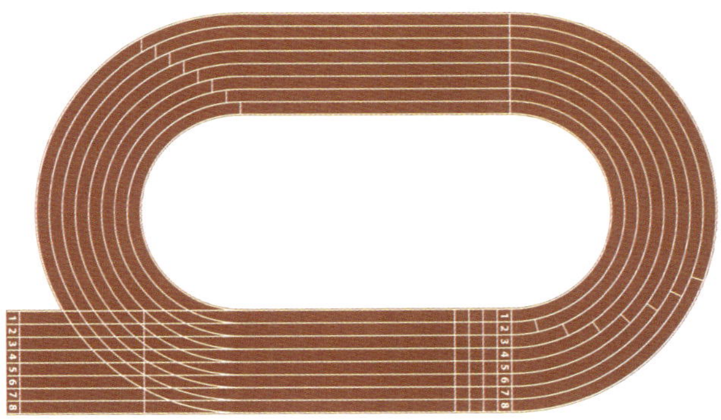

d Kofi's fitness tracker measures the distance around the track in lane 1 as 200 m. He warms up by running $2\frac{1}{2}$ laps. How many metres did he run?

How many 200s in 1,500?

1,500 ÷ 200 =

____

Write as a mixed number.

e The last event of the day is the 1 500 m race. How many laps will these competitors run?

f Marcia says she likes to run in the outside lane (lane 4) when she warms up as it means she runs a little further than the runners in the inside lane. Is she correct? Explain why or why not.

____

____

g (Optional challenge question):
Lane 1 = 200 m
Lane 4 = 207 m
In real races, Lane 4 runners start ahead to be fair.
How far ahead should they start? __________ m

Why? ______________________________

Part 3

Total area = 80 × 80 = ____ m²

Lower primary = $\frac{1}{4}$ × ____ = ____ m²

Sports area = ____ - ____ = ____ m²

Developing a playground

Your school has a square plot of land that the principal decides to use as the main playfield. A piece will be used for the lower primary as swings and slides, and children will have access to the rest of the field to play sports. (Diagram below not drawn to scale).

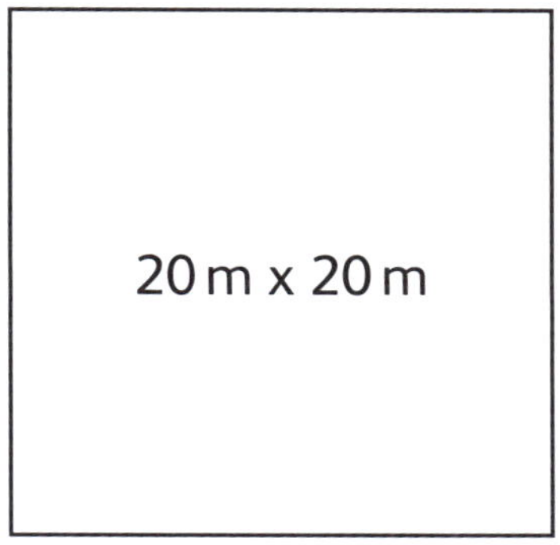

Question 4

Remember, the plot is square.

a The school will fence around the perimeter of the plot. What is the length of the fence?

b The school will allocate $\frac{1}{4}$ of the area to the lower primary. What area of the field will be left for the other grades to play sport?

Part 4

Athletics championships

The Jamaica Teachers' Association and Sagicorp Primary and All/Age school annual track meet will be held at the National Stadium. Many schools from across Jamaica will participate in the sporting event over a three-day period, Thursday to Saturday.

Fifty (50) students participate in their preferred sporting events as seen below:

1 200 m	100 m	200 m	long jump	400 m	800 m	80 m	150 m	ball throw	long jump
80 m	80 m	100 m	long jump	100 m	100 m	100 m	80 m	80 m	1 200 m
400 m	100 m	100 m	high jump	200 m	200 m	200 m	150 m	800 m	1 200 m
1 200 m	100 m	200 m	ball throw	800 m	100 m	1 200 m	150 m	400 m	80 m
80 m	400 m	400 m	ball throw	150 m	100 m	100 m	200 m	400 m	long jump

Question 5

Go through the table one at a time and use a pencil or pen to help you.

a Complete the tally chart to organise the data given above.

Track and field event	Tally	Total number of participants
80 m		
100 m		
150 m		
200 m		
400 m		
800 m		
1 200 m		
Long jump		
High jump		
Ball Throw		
Total		**50**

b In which sporting event did the **least** number of students participate?

____

c Which event has the **most** participants?

____

Remember to:

- Label the x-axis (events).
- Label the y-axis (number of participants).
- Give your graph a title.
- Use a ruler for straight lines.
- Choose a good scale (try 2, 5, or 10).

d On the graph paper provided, draw a suitable graph to show each sporting event and the number of participants. Label your graph clearly so that it is easy to understand.

Write sentences about:

- Which event has the most participants.
- Which event has the fewest participants.
- A pattern you notice.
- Something that surprises

e Write **four** sentences to describe what the graph shows.

____

____

____

____

Reflection

You can get good marks in tasks by clearly showing your working and methods even if you get the final answer incorrect.

Look back at your answers to the task.

- Did you show your working?

____

- Is your working clearly laid out and easy for someone else to understand?

____

- Does your working show that you understood the question and show that you know how to answer it?

____

Next time I will improve my working by:

☐ Writing each step on a separate line

☐ Always including units

☐ Drawing a diagram when it helps

☐ Using a number line for time problems

☐ Checking my work before moving on

The question where I showed the best working was: Q__________ because:

____

Practice paper 4

This task has 4 parts:

- Part 1 has question 1.
- Part 2 has question 2.
- Part 3 has questions 3 and 4.
- Part 4 has question 5.

Answer all the questions.

Write or draw your answer in the spaces provided.

Remember to show all your working.

→ Before you start ...

Read each question.

- Circle any number information in the question. Remember, number information might be given in figures (5, $\frac{1}{2}$ or 1.2) or in words (four, three-quarters, nine kilometres).
- Underline key words in the question. These are clues that tell you what you need to do.
- Highlight any important information that is given in the question. You might need to use this to find the answer.

Helping in the kitchen

Neko, Maria, your mother and Maria's mother all have jobs to do in the kitchen. You are going to help them.

Part 1

Mixing juice

Question 1

Neko has four flavours of fruit juice. The juice comes in bottles and each bottle contains a different amount of juice. He has several bottles of each flavour.

Cherry 200 ml   Mango 300 ml   Papaya 400 ml   Pineapple 500 ml

Neko wants to mix all four flavours, to make **2.5 litres** of mixed-fruit juice. Once Neko opens a bottle, he must use all the juice in it.

Use the information in the diagram.

a Which bottle contains $\frac{1}{2}$ litre of juice?

____

To convert from litres to millilitres you multiply by 1000.

b Which two flavours could you combine to make exactly **0.6 litres** of juice?

____

c How many of each bottle does he need to combine to get **2.5 litres** of juice?

____

____

____

____

Think about how many millilitres will make 1 litre.

Complete the table to show **two different solutions.** Write the number of bottles for each flavour and the amount of each flavour in litres. **For example, bottle X = 0.4 L.**

Flavour	Number of bottles and amount in litres	
	First solution	Second solution
Cherry		
Mango		
Papaya		
Pineapple		
Total		

d Neko decides to make fruit punch for his dad. His dad does not want the pineapple flavour, but will take all the other flavours. Show how Neko can make **2.5 litres** of fruit punch with the other **three** flavours – mango, papaya and cherry. Show two different solutions.

Flavour	Number of bottles and amount in litres	
	First solution	Second solution
Cherry		
Mango		
Papaya		
Total		

Part 2

Measuring liquids

Maria has four measuring jugs in her kitchen. Two of them are empty and two of them have liquid in them. The diagram shows the scale of the four jugs and the amount of liquid in jug **A** and jug **B**.

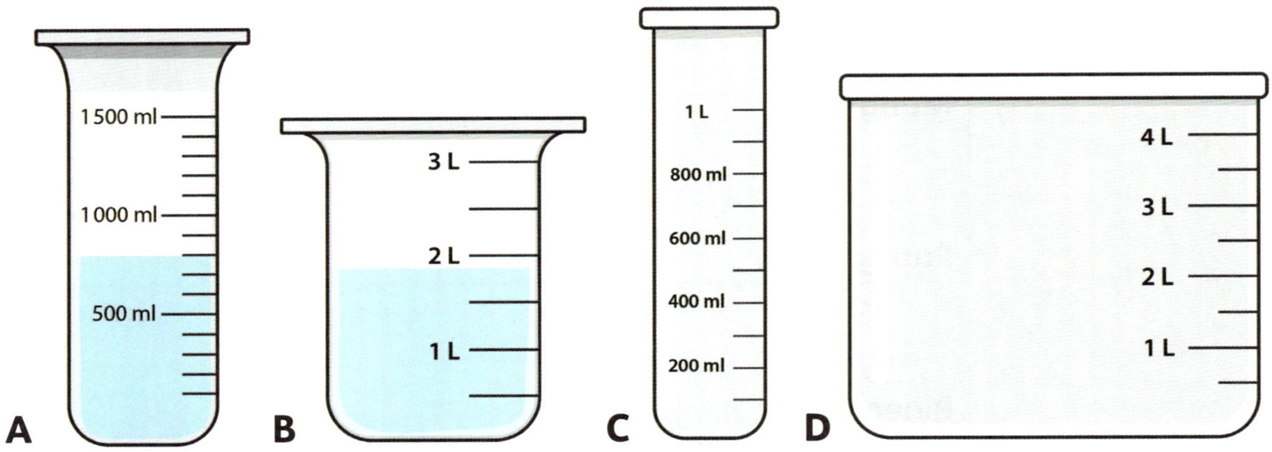

Question 2

Work out what the small divisions on each scale represent before you start.

a How much liquid is in jug **A**?

b Estimate the amount of liquid in jug **B** as accurately as you can. Explain how you decided on that amount.

Draw a line to show the level before you colour in the liquid.

c Maria pours $\frac{1}{2}$ litre of liquid into jug **C**. Shade the diagram to show where the liquid will come to.

d Maria pours the liquids from jugs **A** and **B** into jug **D**. Shade the level of liquid now in jug **D**.

e Circle each measurement below that is equivalent to 1 500 millilitres.

150 centilitre | **0.15 decilitre** | **1.5 litre** | **15 kilolitre**

f Maria weighs 0.35 kilograms of sugar on the kitchen scale. Which arrow shows this weight? Circle the correct arrow.

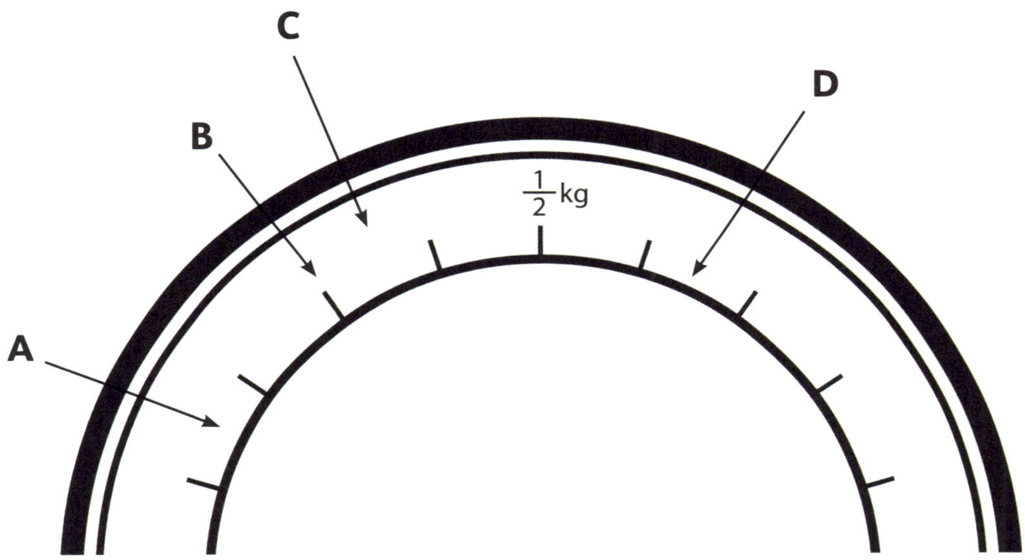

Part 3

What is cooking in the kitchen?

Question 3

You and your mother are in the kitchen. She decides to show you how to cook chicken broth using her granny's recipe.

Granny's chicken broth

$2\frac{2}{5}$ litres of water
$\frac{3}{4}$ kg of yam
500 g chicken breast
250 g pumpkin
45.5 mg salt
45 g noodles
60 ml heavy cream

Remember, we write ratios with a colon between the numbers. For example, a ratio of 4 to 7 is 4:7.

a Granny's recipe used 500 g of chicken breast and 250 g of pumpkin. How much more chicken would be needed to double the amount of pumpkin?

b The recipe uses $\frac{2}{3}$ kilogram of yam and 500 g of chicken breast. Which ingredient is heavier and by how many grams?

These are the instructions for cooking Granny's chicken broth.

Step 1: Boil pumpkin in water with salt.

Step 2: After 40 minutes, add chicken breasts.

Step 3: Wait 25 minutes, then add yam.

Step 4: Simmer for 15 minutes, then add noodles.

Step 5: After ten minutes, add heavy cream.

Step 6: Reduce heat and simmer for 5 minutes, then turn off.

Remember, you can use hops on a number line to solve problems involving time.

c How long does it take to make the chicken broth?

You already know how long it takes. Use your answer from part c.

d One afternoon, mom started making chicken broth at 4:25 p.m. At what time will she be finished?

e You decide to plan your time. Use the instructions to work out the time at which you need to perform each step. Write the times on the lines provided below.

Step 1: Boil pumpkin
Time 4:25 p.m.

Step 2: Add chicken breasts
Time: ____________

Step 3: Add yam
Time: ____________

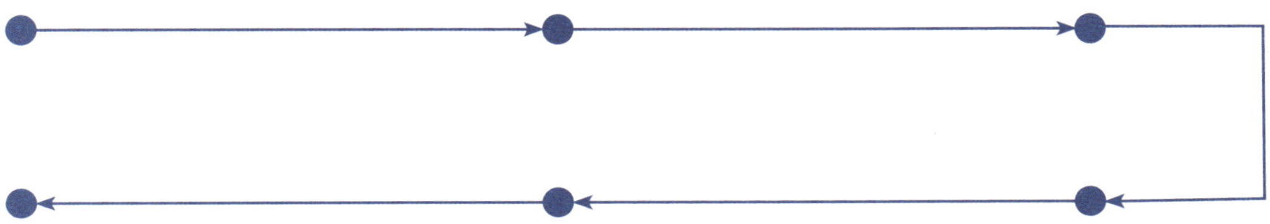

Step 6: Turn off stove
Time: ____________

Step 5: Add heavy cream
Time: ____________

Step 4: Add noodles
Time: ____________

Think carefully about what you are being asked here.

f The recipe makes four (4) servings. However, some friends came over for a short visit and mom had to triple the recipe to have enough for everyone.

i How many people visited the house?

ii How much yam will you need if you triple the recipe?

Question 4

While the chicken broth is cooking, your mother notices that the tiles behind the stove are damaged from heat and steam. She decides this is a good time to retile the area.

Maria's mom decides that the kitchen wall above the stove needs to be retiled. The area to be tiled is 0.16 m high and 4.29 m long.

a The tiles are rectangular, 13 cm x 22 cm. Draw a diagram to show approximately how many tiles will fit across and how many will fit down the wall.

b Do you think they should arrange the tiles horizontally or vertically? Draw both options and explain which looks better.

c The tiles come in boxes of ten and no half or partially full box is sold. How many boxes will they need to buy?

d One tile costs $545. How much will they pay for the boxes of tiles they buy?

Before you start, work out where the line(s) of symmetry will be and complete the pattern.

e Maria likes symmetrical arrangements. She started to sketch some of the tiles she saw at the shop. Complete each tile so that it is symmetrical and draw the line(s) of symmetry.

f (Optional challenge question):
Create your own symmetrical tile pattern in the grid below.
Your pattern must have:
At least TWO lines of symmetry
At least 8 black squares
At least 8 white squares

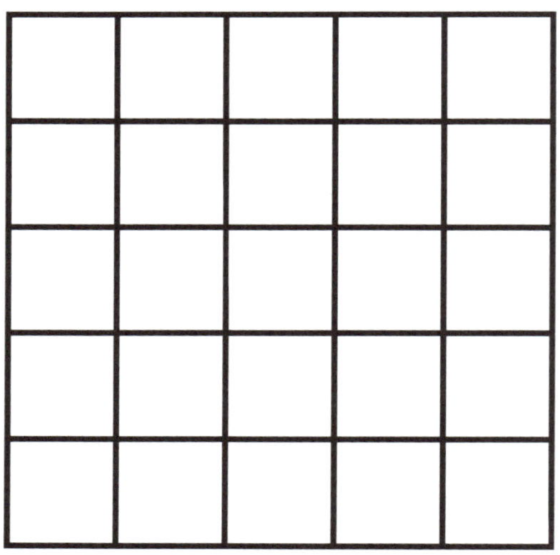

Part 4

Growing your own ingredients

While shopping for ingredients, you see corn seeds on sale. She decides to plant the grains in her backyard. She **only** has a triangular area in the garden available to plant corn. She divides the triangular area into rows. She plants one seed in the first row, three in the second row and five in the third row, and continues increasing the numbers in the same way.

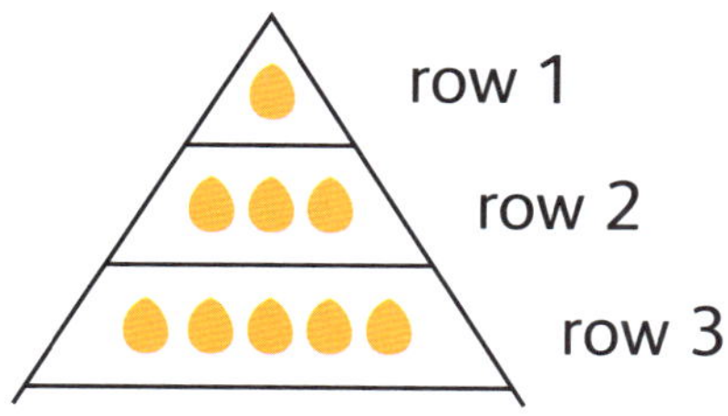

Question 5

a How many corn seeds will there be in the tenth row?

Row 1 has one corn plant, row 2 has three corn plants, row 3 has five corn plants and so on.

b How many corn seeds does she plant altogether in the first ten rows?

How can you find the sum of all the corn seeds in each row?

c Write a rule for working out the number of corn seeds in any row of the plot. Use *n* to represent the row number.

____

____

____

d Use your rule to work out how many seeds will be planted in the 100th row.

e In which row will she plant 39 corn seeds? Show clearly how you work this out.

Reflection

You answered questions in this task.

Use this scale to rank the questions in order from the one you found easiest to the one you found most challenging.

Easiest				Most challenging

Complete these sentences.

- I found question ________ easiest because

______________________________.

- I think question ________ was challenging for me because

______________________________.

Practice paper 5

This task has 3 parts:

- Part 1 has questions 1, 2 and 3.
- Part 2 has question 4.
- Part 3 has question 5.

Always check that you have everything you need before you start.

You will need a calculator, a pencil and an eraser.

Answer all the questions in each part in the spaces provided.

Remember to always show your working out.

→ Before you start ...

Read through all the questions.

Ask yourself: "What am I being asked to do here?"

Jot down your ideas next to each question.

Here are some key words that might help you.

Use a formula Do a calculation
Solve a two-step problem Draw a graph
Show working on a diagram Use patterns
Give an opinion Make decisions Give reasons

Volunteering at the dog shelter

Nadia and Tyrone volunteer to feed dogs at a local dog shelter.

They each work in different sections. You are going to help them to:

- work out how much food different dogs will need
- calculate how long a bag of food will last
- read and make sense of food labels
- work out the amount of artificial turf they will need for the shelter yard
- design a dog kennel and calculate the amount of varnish they will need to cover it.

Part 1

How much food?

Question 1

Nadia feeds the dogs in Section A. These dogs are recovering from operations or illness and they are fed special tinned food. Each dog must get 100 g of food for every 5 kilograms of their mass.

a How much food will Nadia have to give a dog that weighs 15 kilograms?

b What is the rule for working out how much food to give any dog?

Use letters to write a short, clear rule. You will need to use your rule to answer part c.

c Work out how much food each of these dogs should get per day.

Type of dog	Mass of dog	Amount of food
Chihuahua	2.5 kilograms	
Jack Russell	8 kilograms	
Border Collie	22 kilograms	
German Shepherd	39.5 kilograms	

d The tins of dog food each weigh 400 grams. How many tins will Nadia need to feed the four dogs in the table?

Would it be easier for people if she asked for a round number?

e A tin of dog food costs $438.69. Nadia plans to raise funds for the rescue organisation by asking people to sponsor a tin of food. What amount do you think she should ask them to contribute per tin? Give a reason for your answer.

Question 2

Tyrone feeds the dogs in Section B. They eat dry food. He gets these instructions:

Daily food allowance (Split evenly between meals)			
Size of dog	Small (< 10 kilograms)	Medium (10–25 kilograms)	Large (> 25 kilograms)
Amount of food	4% of body weight	3% of body weight	2% of body weight

Read the table carefully before you try to answer the questions.

Think about the steps you need to take to work this out.

a Work out how much food each of these dogs will get at each meal. Round your answers to the nearest 10 grams.

A Great Dane that weighs 59 kilograms and is fed twice a day.

A Maltese poodle that weighs 5 kilograms and is fed three times a day.

A 24-kilogram Border Collie that is fed once a day.

b The food comes in 44-pound bags. Tyrone knows that 1 kilogram = 2.2 pounds. Show how he can work out how many days he can feed the Great Dane from one bag.

Question 3

Read the information on the bags of puppy and adult dog food.

Breakdown Puppy Food:

Crude Protein, minimum	30.0%
Crude Fat, minimum	20.0%
Crude Fibre, maximum	3.5%
Moisture, maximum	10.0%
Omega-3 Fatty Acid, minimum	0.35%
Omega-6 Fatty Acid, minimum	2.50%

Breakdown Adult Dog Food:

Crude Protein, minimum	26.0%
Crude Fat, minimum	11.0%
Crude Fibre, maximum	4.5%
Moisture, maximum	10.0%

a What does puppy food have that adult food does not have?

____

____

What percentage would you use to describe all the ingredients in the food?

b Do you think adult food contains any other ingredients? Explain why or why not.

____

____

____

c What mass of protein would you find in a 15-kilogram bag of puppy food?

____

____

d What is the greatest mass of fibre you would find in a 12-kilogram bag of adult food?

____

____

Nutrition refers to the benefits that we get from the food we eat.

e Explain what the ingredients tell you about the different nutritional needs of puppies and adult dogs.

____

____

____

f (Optional challenge question):

A pet food company wants to create a new 'Junior Dog' food for dogs aged 1–2 years (between puppy and adult).

Based on the nutritional information for puppy food and adult food, design the percentage breakdown for 'Junior Dog' food. It should have:

- More protein than adult food, but less than puppy
- More fat than adult food, but less than puppy
- Similar fibre to adult food

Your Junior Dog Food formula:

Crude Protein: ______________ %

Crude Fat: ______________ %

Crude Fibre: ______________ %

Moisture: ______________ %

Explain your choices. Why did you pick these percentages?

____

Part 2

Shelter space

Question 4

Look at the plan of the animal shelter. The white squares represent the open yard area and the dark squares show the area occupied by the buildings.

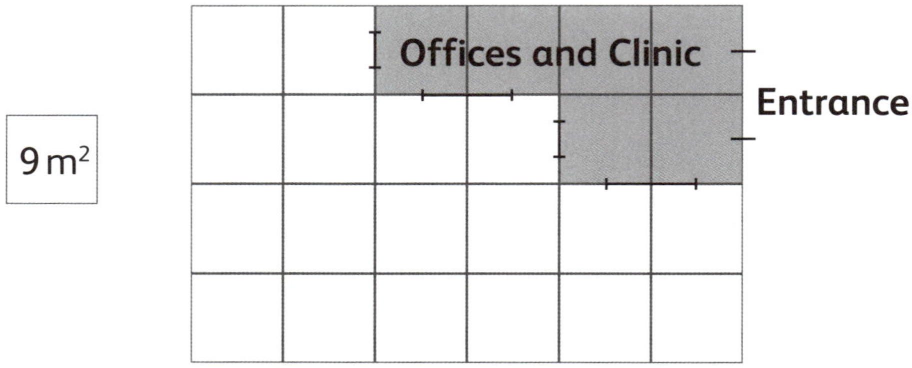

What does the key of the diagram tell you? How long is the side of each square on the diagram?

a What is the area of the building?

You can sketch on the diagram to help you work this out.

b The shelter covers the yard with artificial turf. The turf is sold in rolls. Each roll is 5 m long and the turf is 2 m wide. How many rolls will they need to cover the whole yard?

c The shelter allows a 4 m^2 space per dog. If they are completely full, how many dogs are there?

d A small dog walks around the entire perimeter of the yard. Mark its route on the diagram. How far does it walk?

Part 3

Design a dog kennel

Question 5

Tyrone and Nadia plan to build a simple wooden kennel for their own dog. Their dog is 61 cm tall and 78 cm long from its nose to the base of its tail.

They look online and they find these formulae for working out the size of kennel that they need:

Floor area = (length of dog in cm + 15 cm)^2

Height = height of dog (cm) + 15 cm

Think about the simplest shape you could use for a dog kennel. Remember that Area = L × B, so B = A ÷ L.

a Design and draw a simple dog kennel for their dog. Your diagram should clearly show all the measurements.

They decide to varnish the outside of the kennel to keep it waterproof. Read the information about the varnish before you answer the questions.

Woodlock Waterproof Varnish

1-litre:	J$1 304.70
Coverage:	11–14 m^2 per litre
Drying time:	16 to 18 hours
Time between coats:	minimum 18 hours
First coat:	dilute 30% turps
Second coat:	dilute 5% turps
Third coat:	undiluted

What is the area of each wall and the roof?

b Using your design, work out the total area that they will need to varnish.

c They decide that they will apply three coats of varnish. How many tins will they need to buy to do this?

d How much turps will they need for the first coat?

e Work out the minimum time it will take to apply three coats of varnish and to let the varnish dry completely.

Reflection

Once you have checked your answers to the task, complete these statements about your work.

- I now know

  ______________________________.

- I need to know more about

  ______________________________.

- These things went well

  ______________________________.

- I could do better if I

  ______________________________.

- In my next task I will pay attention to

  ______________________________.

Practice paper 6

This task has 3 parts:

- Part 1 has questions 1 and 2.
- Part 2 has question 3.
- Part 3 has question 4.

Always check that you have everything you need before you start.

You will need a calculator, a pencil and an eraser.

Answer all the questions in each part in the spaces provided.

Remember to always show your working out.

→ Before you start …

When you do a task or test, you can answer the questions in any order.

It is better to do the questions that are easy for you first. That way you have more time to spend on the more challenging parts of the task.

This task has **four** questions.

Read the questions and decide on an order to answer them.

Write your order here:

First, I will do question ________.

Then, I will do question ________.

Next, I will finish question ________.

Lastly, I will do question ________.

Give **one** reason why you chose this order. Write it here.

____

Our food habits

Class 5 are learning about healthy eating and they decide to find out more about the fat content of different foods. You are going to help them to:

- Read data to compare the fat content of different foods.
- Identify foods with a high proportion of fat.
- Compare foods based on their fat content.
- Suggest foods for different dietary requirements.
- Represent data about eating habits graphically.
- Report on research findings about eating habits.

Part 1

What is in our food?

Question 1

The table gives you information about the approximate fat content of some foods.

Food	Mass of food (g)	Mass of fat (g)
Fried egg	45	15
Full fat milk	200 (ml)	7
Banana	150	1
Apple	150	0.1
Lean steak	150	16
Jerk chicken	120	6
Fried chicken	75	12
Beans	200	0.5
Avocado	100	23
Beef burger	215	33
Potato chips	100	37
French fries	175	26
Choc chip cookies	50	5
Cheese	20	8

Use the information from the table to answer these questions.

a Look at the table. Circle the three foods with the highest proportion of fat. Put a star next to the three foods with the lowest proportion of fat.

b What fraction of cheese is fat?

c Which has the lowest proportion of fat: apples or beans? Show how you decided.

d Which is higher in fat: French fries or beef burgers? Show how you decided.

e How much avocado would you need to eat to take in 1 gram of fat?

Think about what you did to solve the problems. That will give you some ideas for your answer.

f Explain why you cannot just look at the grams of fat in the table to work out which foods have the highest proportion of fat.

Question 2

Sondra makes 12 cookies using 100 g of butter (which is 80% fat).

a Draw a diagram to show how the 80 g of fat from the butter is divided among the 12 cookies.

Potato chips are 37% fat.

b Use your diagram to explain which has more fat: 1 cookie or 10 g of potato chips.

c Marcia wants a snack with less than 1 g of fat. Use the table from Question 1 to suggest three snacks she could choose. Explain why each one meets her goal.

Part 2

Planning a healthy diet

Question 3

Dieticians recommend that healthy, active 12-year-olds should take in 1 800–2 200 kilocalories per day. They should eat 65–70 grams of fat per day.

a Suggest a combination of foods from the table in Question 1 that a 12-year-old could include in their daily meals without exceeding the recommended amount of fat.

b One gram of fat contains 9 kilocalories. What proportion of a 12-year-old's daily kilocalories should come from fat?

c Misha finds this information about eggs.

Yolk	White
Kilocalories – 55	Kilocalories – 15
Fat – 4.5 g	Fat – 0 g
Cholesterol – 184 mg	Cholesterol – 0 mg
Carbohydrates – 0.5 g	Carbohydrates – 0 g
Protein – 2.5 g	Protein – 4 g

A whole egg has white and yolk.

i Calculate the fat content of a whole egg as a percentage.

ii Misha says that it is much healthier to eat an omelette made from egg-whites (no yolks) than an omelette made with whole eggs. Use mathematics to support Misha's idea.

If you are asked to use mathematics, it means use numbers or calculations to show that the idea is correct.

d Kai says that you can get all the fat you need in your diet by eating carrots (0.1 gram of fat per medium carrot), broccoli (1 gram of fat per 180 grams) and beans. Explain why his idea is correct but not practical.

e Anish finds this information on a pack of beef patties. Complete the statements about the information.

Nutrition Facts

Serving Size 1 Patty, 5.33 oz (151 g)

Servings Per Container 2

Amount Per Serving

Calories 240 Calories from Fat 110

% Daily Value*

Total Fat 12 g **18%**

i If you eat the whole pack, you will have eaten ________ servings and consumed ________ kilocalories.

ii ________ % of the kilocalories in a serving comes from fat.

ii You'd need to eat ________ patties to get 100% of your daily fat allowance. This is equivalent to ________ g of fat.

Part 3

Changing habits

Question 4

A researcher has investigated this data about the proportion of different types of food that Jamaican families eat.

Type of food (as a percentage of weekly food consumption)	1990	2020
Raw and prepared at home	60%	19%
Ready-made meals and/or packaged food	15%	42%
Takeaways, canteen or bakery goods	7%	31%
Restaurants (eat in) and other	14%	8%

Use the information from the table to answer these questions.

Think about what type of graph you can use to show changes over time like this.

a Design and draw a suitable graph to display this data clearly.

b Comment on how eating habits changed in Jamaica from 1990 to 2020. Suggest reasons for each change.

c What surprised you most about this data? Why?

You are being asked for your own opinion here, but you must give reasons to support your answer.

d Do you think this data reflects your community's eating habits or not? Give clear reasons for your opinion.

Reflection

Give yourself some personal feedback about your work on this task. Use these sentence stems as starters.

- I am already quite good at ______.
- I can get better at dealing with ______.
- These hints will help me improve ______.

Practice paper 7

This task has 3 parts:

- Part 1 has questions 1 and 2.
- Part 2 has questions 3 and 4.
- Part 3 has question 5.

Always check that you have everything you need before you start.

You will need a calculator, a pencil and an eraser.

Answer all the questions in each part in the spaces provided.

Remember to always show your working out.

→ Before you start …

Read through the questions.

Find and highlight all the question words (what, which, where, why and when).

Think about how you can answer each question.

Find and circle these instruction words.

Match	Draw	Calculate	Give a reason
Complete	Predict	Show	Use information

Think about what each word tells you to do.

School events

A local primary school is going to host a dinner and a school concert to help pay for their athletics team to travel to a competition in St Lucia.

You are going to help them by ranking events in the order they need to be completed, solving catering problems and working out costs and income.

Part 1

Organising the dinner

Question 1

Ms Parker is in charge of planning and organising the fundraising dinner. She starts by brainstorming some of the tasks that will need to be done and then writing them on sticky notes.

- ☐ Book the hall at least a month before the event.
- ☐ Work out the final numbers for catering.
- ☐ Send out invitations for the dinner on social media.
- ☐ Sell tickets and collect the money.
- ☐ Order tables and chairs.
- ☐ Pay the deposit for the hall – three weeks before.
- ☐ Pay the balance to rent the hall.
- ☐ Set up the tables and chairs.
- ☐ Organise the delivery of the tables and chairs.

Work in pencil and think before you finalise the answers.

a In what order do the tasks need to be done? Number the sticky notes from 1 to 9 to show the order.

b List two things you thought about when you ordered the events.

____

____

c The dinner will be held on 17 May. It will cost $14 500 to hire the hall for the evening. The school must pay 35% of this amount as a deposit, no longer than three weeks before the event.

i How much is the deposit?

ii What is the latest date that the deposit can be paid?

iii The balance of the cost is due on the day of the event. How much is this?

d The school sells 172 tickets for the dinner. The tickets cost $1 750 each. The food will be donated by local businesses, so the school must only pay for the use of the hall. How much money will they make from the event?

Question 2

There will be 21 VIP guests at the event. They will sit at one long table on the stage. The other tables will be set for eight or nine guests.

Remember, the VIP guests do not sit at these tables.

a Ms Parker works out that she will need two tables of eight. How many tables of nine guests will there be?

b The caterer estimates that each person will drink one glass (300 ml) of water with their meal. How many 2-litre jugs of water will they need for all the guests?

c Each plate will get five vegetable servings (rice and peas, callaloo, okra, fried plantain and salad) and two pieces of meat (chicken or goat). Each person will also get a fruit kebab and three sweet cookies for desert.

i Work out the total number of vegetable servings they will need.

ii The caterer estimates that 75% of the guests will take chicken rather than goat. How many pieces of chicken is that?

iii The sweet cookies come in packets of 12. How many packets will they need to give each person three cookies?

iv Each fruit kebab has one strawberry, one piece of pineapple, three grapes and two pieces of watermelon. The caterer has already made 75 kebabs. How many more pieces of watermelon will he need to make the rest?

v In one fruit bowl, there are three strawberries, five grapes, two pieces of pineapple and two pieces of watermelon. If you picked one fruit without looking, what is the probability that it would be a strawberry?

Part 2

Making choices

Question 3

The students in Class 5 plan to sell cans of soda and packets of chips at the school concert to raise funds. They need to decide how many cartons of soda and boxes of chips to buy so that they do not have many left over.

Highlight the numbers in the information.

a Read the information and decide how many cartons of soda and boxes of chips they should buy. Show how you work this out and give reasons for your answers.

There are seven classes.
Each class has about 28 students.
Usually about 75% of the students come to the concert.
Besides the students, we estimate another 250 people will attend.
The sodas come in cartons of 24.
The chips are packed in boxes of 20.
We estimate that each person will buy one soda and one pack of chips

b When the students go to buy the soda they find that there is a special offer.

Special offer
4 six-packs of Sweet Soda
$1 450

Bargain
25 cans of Cool Cola
$1 460

The best deal means the most soda for the least money.

Which one should they buy to get the best deal? Why?

____

____

Question 4

At the concert, the students sell four flavours of chips (plain, chilli, jerk chicken and sea salt) and three flavours of soda (cola, lime and mango).

Save time by using letters for each flavour instead of writing them out in full.

a How many different combinations of chips and soda are possible? Draw a table or diagram to show how you work this out.

b If you pick a soda and a packet of chips without looking, what is the chance that you will get plain chips with either a cola or lime soda?

c Mai says that if you add another soda flavour, you will get the same number of choices as when you add another chip flavour. Explain why Mai's thinking is incorrect.

You are told here that Mai is incorrect. You have to explain why that is.

Part 3

Cleaning up

Question 5

Ten students are helping to clean the hall and pack up the chairs after the school concert.

a There are 25 rows of 18 chairs and four rows of 15 chairs. If each student carries the same number of chairs to the storeroom, how many will they carry?

b The chairs are stacked in eights. How many stacks of chairs will there be?

c The diagram shows the height of a stack of three chairs. How high will a stack of eight chairs be?

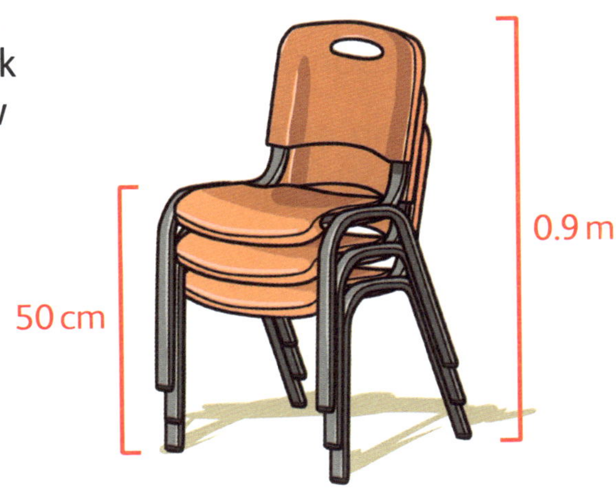

Write number sentences and solve them to show your working.

e The students also stack the basins that were filled with ice and used to cool the sodas.

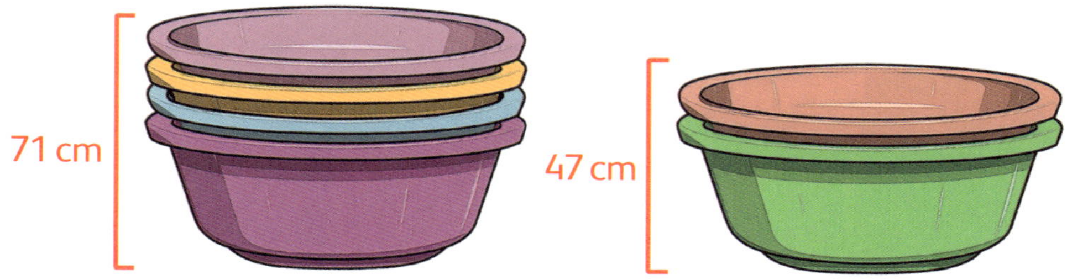

How tall is one of these basins? Explain clearly how you work this out.

Reflection

You already know that thinking about your mistakes can help you improve.

Find any mistakes in your work.

For each one, decide whether it was:

- **careless** (writing the wrong number, not following instructions, forgetting to give reasons, not showing steps in calculation)
- **calculation error** (miscalculating a value using the basic operations).

Write down **three** suggestions to help you avoid making these mistakes in the future.

Practice paper 8

This task has 3 parts:

- Part 1 has question 1.
- Part 2 has questions 2, 3 and 4.
- Part 3 has question 5.

Always check that you have everything you need before you start.

You will need a calculator, a pencil and an eraser.

Answer all the questions in each part in the spaces provided.

Remember to always show your working out.

→ Before you start …

1 Complete this checklist before you start the task.

☐ I have read the instructions and I have the things I need.

☐ I have read through the questions.

☐ I have underlined or highlighted important information.

2 Look at the questions carefully. Use this table to sort them into three different types of answers.

Simple calculations and answers	Thinking and giving reasons	Many things to work out, more than one possible answer

Keeping animals at school

Class 5 are studying animals in science. They plan to keep some animals in the yard at school so they can learn what the animals need, how they behave and how to care for them. You are going to help them by:

- interpreting and making sense of data
- working with diagrams and measurements
- drawing plans to show pens for tortoises
- working out how much food different animals need
- calculating the costs of keeping a bird at school.

Part 1

Choosing an animal to keep at school

The students decide to do a survey to find out which animals people keep in their yard at home. They want to use the data to choose an animal that is familiar to some of the students.

Question 1

Alicia's group collects data by giving everyone a piece of paper and asking them to write down which animals they keep at home. These are the results they get.

Four chickens and a cat	None	A tortoise, a dog and some fish	Six chickens and a terrier
We don't have any	Dogs	Cats and 12 chickens	Three white chickens and a brown chicken
Some lizards	A dog, a grey cat and a parrot	A donkey and seven chickens	A turtle, I think
Some wild birds	Two cats	Three dogs and a chicken	12 fish and a ginger cat
Two rabbits and four big white chickens	My cousin down the road has a dog	I think there might be mice	We had seven chickens, but there are only five now
None	Four rabbits, but one is sick	Two brown chickens and a white chicken	A very skinny goat and two chickens

Use a pencil and work lightly in case you change your mind.

You do not need to use all the rows.

a Some of the information is not relevant or necessary. Cross out the bits of information that are not needed.

b Why is it difficult to make sense of this data? Give **one** reason.

____

c Complete this tally chart to organise the data.

Type of animal	Tally	Total

d According to your chart, which animal is kept by most people? ______________________

e Which animal do you think is most familiar to the students? Explain why.

____

Part 2

Chickens and two tortoises

Question 2

The students in Grade 5 class decide that they will keep chickens at school. One of the parents builds them a chicken coop. The chicken coop has a closed shed for the chickens to nest and lay eggs at one side. It has an outside fenced area for the chickens to walk around which is attached to the shed.

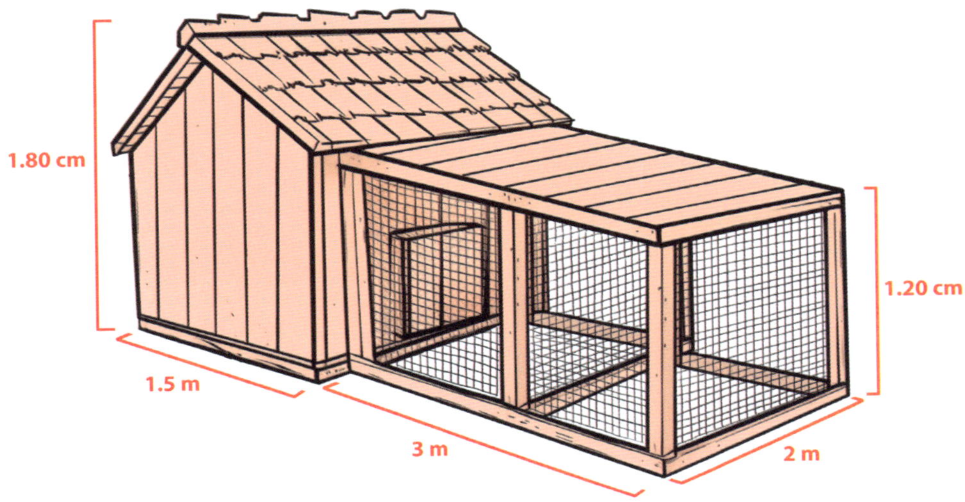

a How much taller is the shed than the outside area?

b The students find out that they need approximately half a square metre of outside space for each chicken. What is the greatest number of chickens they can keep in this coop?

Remember, you can write on the diagram if it will help you.

c Mr Richardson bought 5 m long strips of wood to build the frame for the outside area. How many strips did he need altogether?

d How many square metres of fencing are there on the outside area? Show how you work this out.

Question 3

The students also decide to keep two tortoises. The tortoises will be kept in a pen with a low fence around it. They have 18 metres of fencing.

You could draw a labelled diagram to support your answer.

a If the students make a rectangular pen, what are some possible lengths for the sides? Sketch two rectangles that you think would be a good size and shape for the tortoises. Write the lengths of the sides on your diagrams.

b Is it possible to make a square pen with 18 metres of fencing? Support your answer.

Question 4

Alicia finds this information about feeding tortoises.

Mass of tortoise	Calories needed (per feed)	Fat (g)	Protein (g)	Carbohydrates (g)
1 kg	32	3	13	48
5 kg	214	10	43	162
* Feed daily in the morning, 5/7 days per week.				

a How many calories does a 5-kg tortoise need each day?

Remember, you can write on the diagram if it helps you.

b What does feed $\frac{5}{7}$ days per week mean?

c How many grams of fat and protein does a 1-kg tortoise need per day?

d Alicia says you cannot feed different tortoises from the same bag of tortoise food because they need different amounts of fat, protein and carbohydrates. Use mathematics to show why Alicia's thinking is incorrect.

e If the students feed the tortoises for seven weeks, how many calories will the 1-kilogram tortoise need in that time?

f The table shows that both tortoises need fat, protein, and carbohydrates. Create a simple diagram or chart to show the proportions each tortoise needs. What do you notice about the pattern?

Part 3

Budgeting for a budgie

Question 5

Class 5 are also going to get a budgie to keep in the science laboratory. The school has a budget of $10 000 for the budgie and the things it will need.

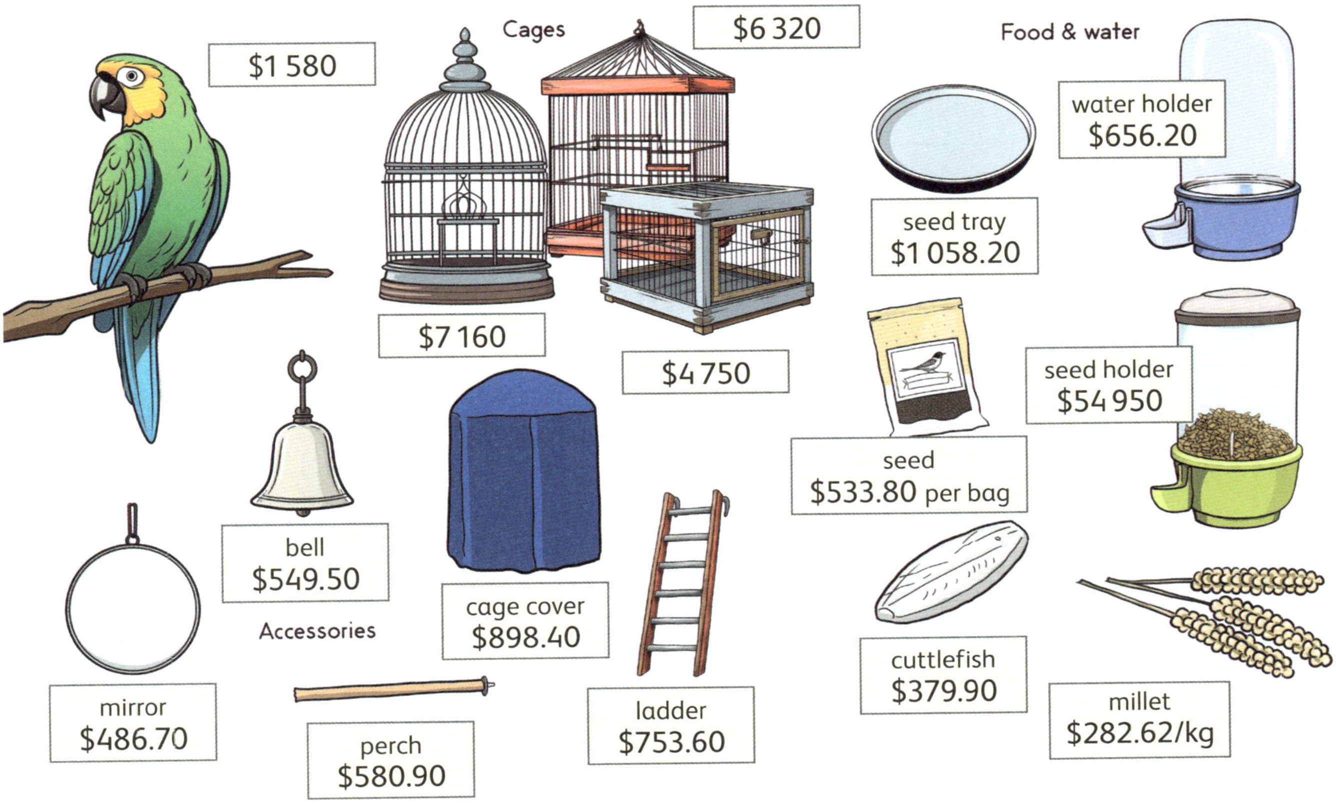

List the costs and then add them up.

a At the start of the term, the school orders and pays for one budgie, the cheapest cage, two bags of seed and 500 grams of millet. How much will this cost?

What is a quick way to find 50%?

b The store offers the school 50% discount on any other items they buy. The school buys a perch, a water holder and a seed holder. How much money will they have left in their budget after this?

c If the school also buys a ladder, a seed tray and a cover for the cage (all with a 50% discount), how many bags of seed will they be able to buy?

d The Science Department will pay for bird seed for the remaining four months of the year. If a bag of seed lasts two weeks, how much money should the Science Department set aside to feed the budgie for the rest of the year?

Reflection

Kofi says, “the best way to learn something is to do it”.

- What do you think he means?

- How can doing practice tasks help you learn more about tasks? Write **two** ideas.

- How do you learn math? Write down **two** things you do.

Practice paper 9

This task has 3 parts:

- Part 1 has question 1.
- Part 2 has questions 2 and 3.
- Part 3 has question 4.

Always check that you have everything you need before you start.

You will need a calculator, a pencil and an eraser.

Answer all the questions in each part in the spaces provided.

Remember to always show your working out.

→ Before you start …

In this task, you will need to find information in diagrams and tables to answer the questions.

Find all the diagrams and all the tables in the task.

- Highlight or circle all the mathematical information in each diagram.
- Underline the headings in the tables. Make sure you understand how to read the information in the table.

Water

Class 5 heard a news report about climate change and how it could affect the amount of rain that falls in Jamaica. They decided to find out more about the importance of saving water at home and school, especially in the dry season.

In this task, you will do calculations using percentages and measurements. You will represent information on a diagram, work out how much water is in a pond and how much is lost by evaporation. You will design an efficient irrigation system and work with data to answer questions about water usage.

Part 1

Water for life

Question 1

The students find this information about what percentage of a human's body weight is water.

Age	0–5	6–20	21–35	36–50	51–80
Percentage of body weight that is water	80	70	60	55	50

a Write the proportion of water in your body as a fraction in its simplest form.

____

b What happens to the proportion of water in your body as you get older?

____

Think about how the scale at the side of each figure can help you.

c The students decide to show some of the information on a chart. Shade the proportion of each body that would be water.

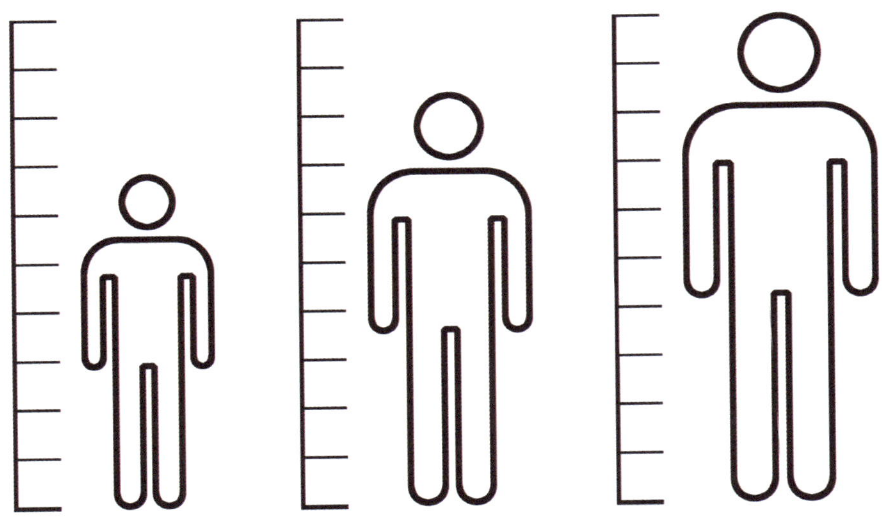

Toddler (3) Teenager (13) Adult (28)

d The students learn that 1 litre of water has a mass of 1 kilogram. If a 12-year-old student weighs 45 kilograms, what is the mass of the water in their body?

e Tonille's baby sister is four. She weighs 16 kilograms. Tonille says that 12 kilograms of her sister's weight is water. Is she correct? Show how you decide.

You have to decide if Tonille is correct or not. How can you use math to do that?

f When you are active, you lose about 9% of the water in your body in a day by sweating. David is very active. He is nearly 14 and he weighs 52 kilograms. How much water should he drink in a day to replace the amount he loses by sweating? Show clearly how you work this out.

Part 2

Storing and using water

Question 2

The students are going hiking in the Blue Mountains to learn more about how farmers in the area collect and store water. Their teacher tells them that they must carry water with them to drink.

a Five students fill up these water bottles. What is the total mass of the water?

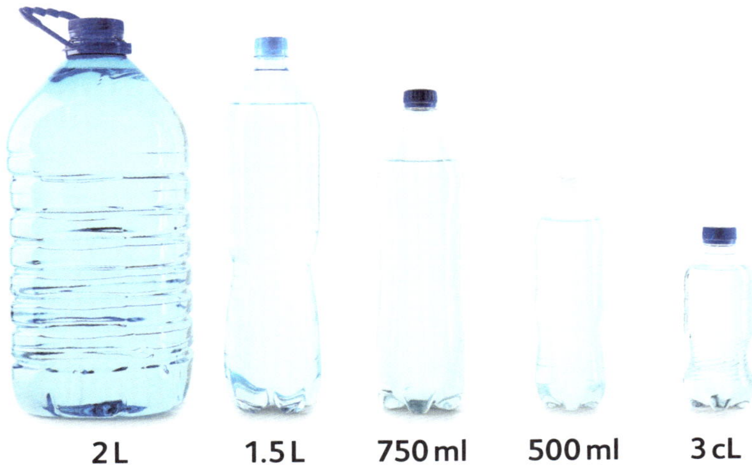

b What is the mean amount of water per student?

volume = length × width × height

c Mr Stocks shows them one of the ponds that he has made to collect and store water. He dug a hole in the ground and lined it with plastic to make the pond. Mr Stocks tells them that there are 1 000 litres in every cubic metre of water. The pond is 6 m long, 3 m wide and 0.75 m deep. How many litres of water can it hold?

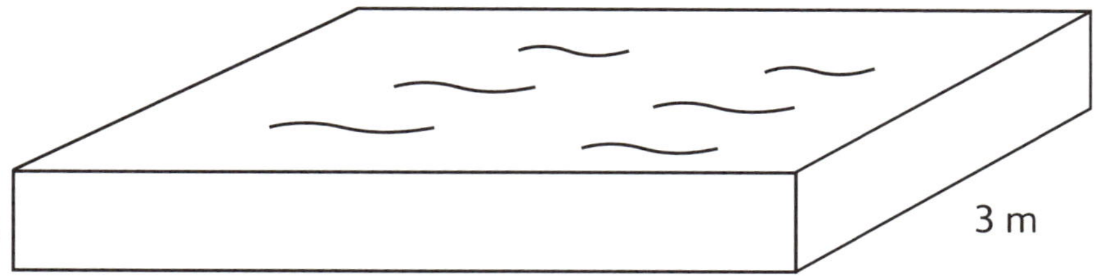

d If the pond is half full, what is the mass of the water in it?

e Mr Stocks tells them that the water evaporates quite quickly in the dry season. The water level today is 45 cm. He tells them that the water was 50 cm deep five days ago.

i Express the evaporation as a rate per day.

ii If no water flows into the pond, how many days will it take for all the water to evaporate at that rate?

Would a sketch help?

iii If the water level drops 10 cm, how many litres of water have been lost to evaporation?

Question 3

Mr Stocks explains that he pumps water from the ponds to water his crops. He has drawn this plan of his piped irrigation system.

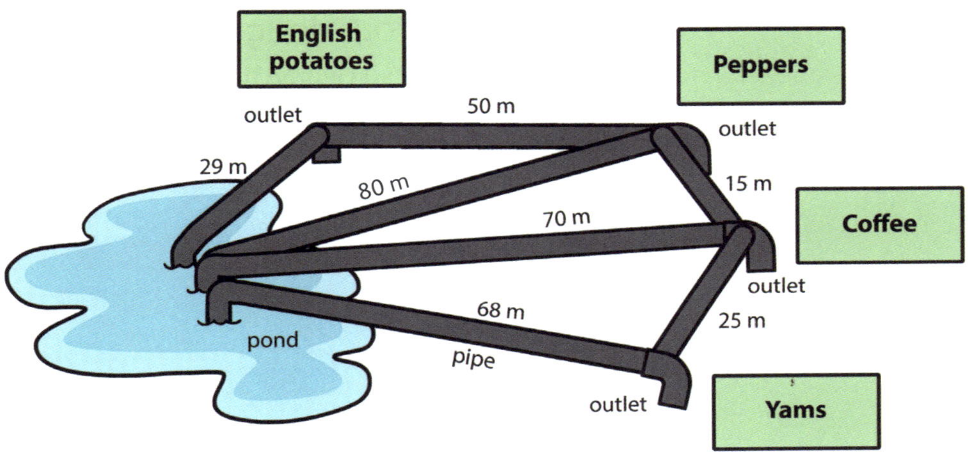

Mrs Stocks says that she thinks this system uses too many pipes.

Work on the diagram above to plan your design before you draw it on the blank diagram on the next page.

a Redesign the irrigation system to show how Mr Stocks could get water to all his crops using the shortest possible length of pipe.

Draw your design on the diagram and include pipe lengths.

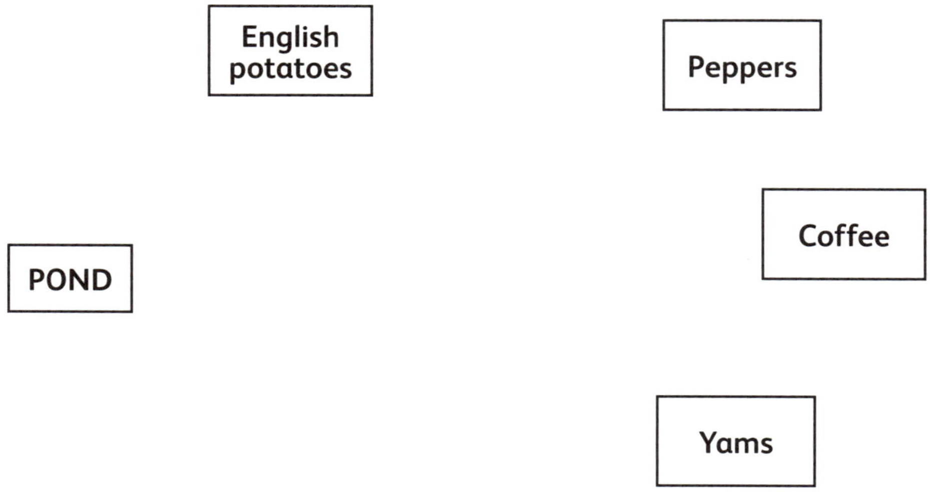

b Compare your design with the original plan. State **two** advantages that your design has over the one on the plan.

Part 3

Using water

The table shows how much water is needed (on average) for some different activities.

Activity	Water used (litres)	Activity	Water used (litres)
Cooking and drinking	30	Bathing	140
Washing hands (2 minutes)	18	Washing clothes	110
Brushing teeth (tap running for 1 minute)	5	Flushing toilet (low flush)	15 5
Showering (5 minutes)	45	Washing dishes	42

Question 4

a How much water do you save if you only shower for four minutes?

b If someone flushes a low flush toilet five times, how much water do they use?

c Alison washes three loads of clothes. How much water does that use?

d John washes dishes twice a day. How much water does that use?

e Sarah washes her hands five times a day. How much water can she save if she runs the tap for $1\frac{1}{2}$ minute each time, instead of 2 minutes?

f A hotel decides to remove all the bathtubs and only have showers. They also display notices in the bathrooms to ask people to use their towels for more than one day. Explain in detail how this can help the hotel save water and money.

g (Optional challenge question)
Making products uses water too:

1 cotton t-shirt = 2,700 litres of water

1 pair of jeans = 10,000 litres of water

1 smartphone = 13,000 litres of water

1 hamburger = 2,400 litres of water

A family buys 3 t-shirts, 2 pairs of jeans, and 4 hamburgers in one month. Calculate the total 'hidden' water used.

Reflection

- Make a list of **three** things in this task that you did well.

- What could you do better?

- Write down **one** thing that you learned in another task that helped you here.

Practice paper 10

This task has 3 parts:

- Part 1 has question 1.
- Part 2 has questions 2 and 3.
- Part 3 has question 4.

Always check that you have everything you need before you start.

You will need a calculator, a pencil and an eraser.

Answer all the questions in each part in the spaces provided.

Remember to always show your working out.

→ Before you start …

You have learned a lot about doing tasks in this book.

Write down **three** important things that you should always do before you start a task.

1 ______________________________

2 ______________________________

3 ______________________________

Why is it important to do these things?

____

____

Carnival time

Carnival is coming and everyone is excited, but there are plans to make first.

Part 1

Parade routes

Question 1

The hand-drawn map shows the planned route for two different carnival events: the Road March and the J'ouvert route. The map has been drawn at a scale of approximately 1 cm to 280 m.

Do not just guess. Use the map to help you decide whether the statement is true or false.

a Write true or false next to each statement.

The Road March starts and ends at the same place.

____

The Road March route goes past the Bob Marley Museum. __________

The J'ouvert route is longer than the Road March route. __________

Both routes end in Oxford Road. __________

Neither route goes down Halfway Tree Road.

____

b Write a clear set of directions for someone who wants to follow the J'ouvert route.

c An estimated 5 000 people will take part in the Road March. A quarter of the people have already passed the Terra Nova Hotel. How many more people will march past?

Remember, it is faster to use letters than to write out the names.

d Four Mas Bands are going to play at Emancipation Park in the morning. They will play one after the other. The bands are: Funky Flamingo, Jolly Jazz, Banging Beats and Reggae Riot. You have been asked to work out the order in which the bands will play. How many possible ways are there to do this?

e Use the scale of the map to estimate the length of the Road March route. Give your answer in kilometres.

Part 2

Budgeting

Question 2

Ziggy lives in the USA. He and two friends want to visit Jamaica to take part in carnival. Ziggy does some research online to find out about costs so they can budget for their trip. This is what Ziggy finds out:

Dates of carnival:	12–18 April
Flights to Jamaica:	Around $300 per person
Accommodation:	Ranges from $120 to $800 per night
Car rental:	$320 per week
Tickets for fetes:	$50–$125
Costumes:	Average for men $500, T-shirt section of Mas Bands $250
Spending money:	$500 per person per week (food, drinks, local tours)

a Ziggy wants to be in Jamaica at least two days before carnival starts. When is the latest they can arrive in Jamaica?

____

b If they stay for five days after the carnival ends, on what date will they leave Jamaica?

____

Think about what mid-range means. How can you work out what that is?

c Ziggy and his friends want mid-range accommodation. What will that cost per night?

d They decide that the most they can spend on fete tickets is $450 per person.

i What is the greatest number of fetes they can attend?

ii If they attend only the most expensive fetes, how many tickets can they afford?

e They will share the cost of hiring one car for the time they are in Jamaica. How much will they each pay?

Question 3

Use the information that Ziggy found and your answers to Question 2 to draw up a budget that shows how much each person will need to spend if they go on this trip.

Part 3

Post-carnival activities

Question 4

After the carnival, Ziggy and his friends decide to visit Dunn's River Falls. They find these directions from their location in Downtown, Kingston, in the mapping app.

Down Town
Kingston, Jamaica

> Drive from Port Royal St, Marcus Garvey Dr, Spanish Town Rd and Mandela Hwy to St Catherine Parish

22 minutes (16.9 km)

> Follow Hwy 2000 to St Ann Parish

46 minutes (66.4 km)

> Drive to your destination

5 minutes (3.5 km)

The World Famous Dunn's River Falls & Park
Ocho Rios, Jamaica

Highlight the numbers in the information.

a How far is the journey?

____

____

____

____

____

b How long does the mapping app say it will take them to get there?

____

____

____

c You can work out speed in kilometres per hour using the formula: Speed (km/hour) = Distance (km) ÷ Time (hours). What is the average speed for this journey according to the mapping app?

d The website for Dunn's Falls says that it takes an average of 1 hour and 30 minutes to climb the falls. The falls are open from 8 a.m. to 4:30 p.m. on the day they will visit. Ziggy and his friends want to climb down the falls and spend at least an hour on the beach afterwards. They want to be back in Kingston by 4 p.m. so they can meet up with some friends they met at carnival. When do you think they should leave Downtown to travel to the falls? Give a reason for your answer.

e These are the entry prices for the falls.

Ticket prices

	Child (4–12)	Adult
Jamaicans	J$500	J$1 000 *must show ID
All other visitors	US $17	US $25

Ziggy and his friends do not have Jamaican ID documents. What will the tickets cost them?

f The exchange rate on the day of their visit is US$1:J$155.

i How much less do Jamaican adult tickets cost compared with those for other visitors?

ii Why do you think this is the case?

Reflection

- What did you find most difficult or confusing in this task?

- What did you do about it?

- How did this help you?